D1559903

Waste Disposal in Academic Institutions

edited by

James A. Kaufman

Library of Congress Cataloging-in-Publication Data

Waste disposal in academic institutions / edited by James A. Kaufman.

p. cm.

Papers presented at a symposium held at the Third Chemical Congress of North America in Toronto, Canada, 1988.

ISBN 0-87371-256-0

1. Chemical laboratories—Waste disposal—Congresses. I. Kaufman, James A. (James Aks), 1943- . II. Chemical Congress of North America (3rd : 1988 : Toronto, Canada)

QD64.W37 1990

363.72'87'02437—dc20 90-5400

CIP

Second Printing 1990

LEWIS PUBLISHERS, INC.
121 South Main Street, Chelsea, Michigan 48118

PRINTED IN THE UNITED STATES OF AMERICA

Preface

Hazardous waste disposal problems confronted by academic institutions have reached critical proportions today. Schools, colleges, and universities are faced with the harsh reality of compliance with regulations that ten years ago did not exist; in many cases, their administrators are not prepared to pay for this compliance.

This book, containing the collected papers from the symposium "Waste Disposal in Academic Institutions" held at the Third Chemical Congress of North America in Toronto in 1988, offers specific, practical, and cost-effective solutions to these problems. In addition, it will help significantly to clarify the legal requirements placed on both secondary and post-secondary institutions.

Waste Disposal in Academic Institutions is intended not only for academic institutions but small businesses as well. As small quantity generators and conditionally excluded small quantity generators, secondary schools, colleges, universities, and small businesses will identify with the problems and solutions presented by the contributors.

I wish to thank the contributors to this book not only for their efforts in presenting papers at the symposium in Toronto, but also for preparing their manuscripts for publication. I also wish to acknowledge the American Chemical Society and its Division of Chemical Health and Safety (DCHAS) for support and the Chemical Institute of Canada for hosting the Congress. Special thanks are due to DCHAS program chairperson Patricia Redden for inviting the symposium to Toronto; I owed her a big favor and could not rightly refuse.

James A. Kaufman is Professor of Chemistry at Curry College in Milton, Massachusetts, and a laboratory safety consultant. He received his BA in chemistry from Tufts University and his PhD in organic chemistry from Worcester Polytechnic Institute (WPI).

After two years as a post-doctoral student in the WPI Chemical Engineering Department converting garbage into fuel oil, Dr. Kaufman joined Dow Chemical's New England Research Laboratory as a Process Research Chemist. At Dow, he became increasingly involved in laboratory safety, authoring a pamphlet, "Laboratory Safety Guidelines." Nearly a half million copies of this widely requested and reprinted brochure are now in circulation. In 1986, with Dr. Kaufman's assistance, Dow revised the "Guidelines" and sent copies to 10,000 high schools across the country.

At Curry College, Dr. Kaufman is the founder and director of the Laboratory Safety Workshop, a national center for training personnel in laboratory safety methods and for distributing information on laboratory safety. The center provides instructional programs, audio-visual materials, and a newsletter for science teachers. The Workshop is supported by grants from companies, foundations, and professional societies.

As a laboratory safety consultant, Dr. Kaufman conducts seminars, inspections, and in-service workshops for schools, colleges, universities, and industry. He also provides advice on facility design and editorial commentary on laboratory texts.

Dr. Kaufman is a member of the American Chemical Society (ACS) Council Committee on Chemical Safety and is a former chairman of the 2500-member ACS Division of Chemical Health and Safety. He is chairman of the Health and Safety Committee of the ACS Northeastern Section, and is the author-narrator of the ACS Audio Course on Laboratory Safety.

Introduction

The Third Chemical Congress of North America was held in Toronto, Ontario in June 1988. Members from the chemical societies of Mexico, Canada, and the United States attended the one-week conference.

The American Chemical Society (ACS) Division of Chemical Health and Safety and the Curry College Laboratory Safety Workshop organized a symposium called "Waste Disposal in Academic Institutions" that was presented at the conference. This book contains the papers given at the symposium, organized into five sections dealing with various aspects of waste disposal problems faced by academic institutions. In addition to those papers from the symposium, two related papers (by Joyce A. Kilby et al. and Russell W. Phifer) presented at a subsequent ACS national meeting have been included.

Over the past nine years, the Laboratory Safety Workshop has been providing training programs for science teachers. Nearly 10,000 have attended the sessions. Without a doubt, waste disposal has been the topic most discussed.

Teachers want to understand the requirements of the laws of their state and the federal government. They want help in convincing their administrations that significant sums need to be spent to hire responsible contractors. With work loads already too high, they wonder where they will find the time to address these issues. They want practical suggestions for dealing with their chemical wastes.

These symposium proceedings address all of these issues and many more. The 14 chapters cover a wide range of waste disposal issues. They present the vast experience of school, college, and university science teachers as well as the efforts of federal and private companies in dealing with waste disposal problems.

Section I, Federal Regulations and the Problems of Academic Institutions, summarizes the U.S. Environmental Protection Agency (EPA) hazardous waste requirements as they apply to academic institutions. It discusses the limitations of these regulations and describes problems faced by secondary schools.

Section II, Academic Waste Disposal Programs, provides three views of waste disposal program development. The section begins with the often painful experiences of a small secondary school in Connecticut, describes a successful program to conduct a one-time cleanout of Illinois secondary schools, and finally, provides an overview of an ideal academic waste disposal program.

Every academic laboratory is faced with the problem of samples and purchased chemicals that have no identity. Section III, Identification of Unknown Chemicals, deals with this problem. The approaches in these chapters can save many chemistry departments thousands of dollars.

Once the unknown chemicals have been identified, they and other unwanted hazardous wastes can be treated in-house to reduce or eliminate hazards. The chapters in Section IV, Methods for Treating and Handling Wastes, illustrate how waste disposal problems can be solved in laboratories or universities.

Section V, Waste Disposal Practices, describes the process of lab packing and the benefits of recycling through waste exchange programs, comparing some of the approaches to waste disposal taken around the country.

As symposium organizer and chairman, I enjoyed having the opportunity to work with each of the author/presenters. I learned much about the disposal of hazardous wastes from their papers and presentations. Their experiences can save considerable time and expense, and we can all benefit from their good ideas.

What about your good ideas for disposing of hazardous wastes? Send them to the Laboratory Safety Workshop at Curry College (Milton, MA 02186) for inclusion in the "Good Ideas" section of the newsletter, *Speaking of Safety*.

For more information on waste disposal, readers are referred to these publications:

1. "Less is Better," "RCRA and Laboratories," and "Hazardous Waste Management" (pamphlets), The American Chemical Society, Office of Federal Regulatory Programs, 1155 16th Street NW, Washington, DC 20036. Single copies of each of these are available without charge.
2. *Prudent Practices for Disposal of Chemicals from Laboratories*, National Academy Press, Division of the National Academy of Sciences, 2101 Constitution Avenue NW, Washington, DC 20418, 1983.
3. *Handbook of Hazardous Waste Management for Small Quantity Generators*, R. W. Phifer and W. R. McTigue, Jr., Lewis Publishers, Inc., 121 South Main Street, Chelsea, MI 48118, 1988.
4. *Hazardous Waste Management at Educational Institutions*, National Association of College and University Business Officers, One Dupont Circle, Washington, DC 20036, 1987.

Contents

SECTION IV

METHODS FOR TREATING AND HANDLING WASTES

SECTION V

WASTE DISPOSAL PRACTICES

Waste Disposal in Academic Institutions

SECTION I

Federal Regulations and the Problems of Academic Institutions

CHAPTER 1

How to Establish an Academic Laboratory Waste Management Program

Antony C. Wilbraham

INTRODUCTION

In 1976 the U.S. Congress passed the Resource Conservation and Recovery Act affectionately known as RCRA. With the passage of the act, the U.S. Environmental Protection Agency (EPA) was directed to implement a program that would "—protect human health and the environment from improper hazardous waste management." In May 1980 the EPA published the first set of RCRA regulations. These regulations focused on large companies, specifically those producing more than 1000 kg (2200 lb or about five 55-gal drums) of hazardous waste per month. The regulations required the use of manifests and record-keeping systems to track hazardous wastes from their generation to final disposal. This procedure soon became known as the "cradle-to-grave" approach.

In November 1984 the Hazardous and Solid Wastes Amendments to RCRA were signed into law. With these amendments, the EPA was directed to include small quantity generators in the hazardous waste regulatory system. A small quantity generator is a business or facility that produces between 100 kg (about half of a 55-gal drum) and 1000 kg of hazardous waste per month. Many colleges and universities were now subject to the regulations. The following are important regulatory changes and effective dates:

March 1986:	EPA issued the final regulations for small quantity generators.
September 1986:	Most of the new rules and regulations for small quantity generators became effective. From this date, noncompliance with the regulations could result in fines and/or legal action.
February 1987:	A limit was set on the amount of liquid waste in containers that were destined for landfill disposal.
March 1987:	Small quantity generators were required to have a RCRA permit if they (1) stored waste for more than 6 months (2)

treated waste on their property or (3) disposed of waste on their property.

June 1990: By this date there will be a total ban on land disposal. Obviously this ban will have a marked effect on how hazardous waste producers, including academic institutions, conduct their business.

CODE OF FEDERAL REGULATIONS

The regulations are contained in a series of volumes known as the *Code of Federal Regulations* or CFR. The *Code* is divided into 50 titles that represent broad areas subject to Federal Regulation; Title 40 (40 CFR) deals with "Protection of the Environment." Each title is divided into chapters that usually bear the name of the issuing agency; Title 40, Chapter 1 is the Environmental Protection Agency. Each chapter is further subdivided into parts covering specific regulatory areas. More specifically, to find the EPA Hazardous Waste Regulations refer to 40 CFR Parts 260–266; Part 261 gives definitions and Part 263 provides rules for transport of hazardous waste. Each volume of the Code is revised at least once each year.

The regulations for the Occupational Safety and Health Act (OSH Act) are given at 29 CFR and those for the U.S. Department of Transportation (DOT) are given at 49 CFR. Hazardous Waste Transportation is detailed at 49 CFR Parts 172–179.

For questions related to the regulations, RCRA has a toll-free hotline, 1-800-424-9346. In addition copies of 40 CFR 190–399, which contains Parts 260–266, are available from the Government Printing Office. Write or phone:

Superintendent of Documents
Government Printing Office
Washington, DC 20402
(Tel. 202-783-3288)

In Canada, contact the provincial office listed in Appendix I.

WHO ARE THE SMALL QUANTITY GENERATORS?

More than half of the small quantity generators fall into one of these five categories: vehicle maintenance workshops, manufacturing and metal finishing plants, printing shops, photographic laboratories, and laundry and dry cleaners. The remainder include wood preserving, analytical and clinical laboratories, construction operations, pesticide applicators, and academic science laboratories of schools, colleges and universities.

Typical hazardous wastes from these generators include spent solvents and chemicals, chemical wastes from manufacturing processes, discarded chemical products, empty chemical containers, chemical spill residues, used lead-acid

Table I. Some Examples from the F-List: Hazardous Wastes from Non-Specific Sources

EPA ID Number Rating	Description	Hazard rating
F001	spent halogenated degreasing solvents, trichlorethylene etc.	(T)
F002[a]	spent halogenated solvents	(T)
F003[a]	spent nonhalogenated solvents, xylene, acetone, ethyl ether	(I)
F004[a]	spent nonhalogenated solvents, benzene, carbon disulfide	(I, T)

Note: See Chapter 1, Appendix II for the entire list.
[a]Waste solvents from organic chemistry laboratories are usually F002, F003, or F004 wastes.

batteries, used engine oil, and a wide range of exotic chemicals from research laboratories.

Academic institutions also generate wastes through their vehicle maintenance workshops, printing operations, photographic services, etc. It is easy for an academic institution to become a small quantity generator.

WHAT IS A HAZARDOUS WASTE?

Basically, a waste is something unwanted. In a broad sense, it is usually a solid, liquid, or a contained gas. It is the responsibility of the generator to determine if the waste is nonhazardous, hazardous, or acutely hazardous. The waste is regarded as a hazardous or acutely hazardous waste if, through improper handling, it can cause injury or death, or can damage or pollute the environment. A hazardous waste is regulated by federal, state, public health, and environmental safety laws.

To determine if a waste is hazardous, it may be necessary to refer to the *Code of Federal Regulations* at 40 CFR Part 261, "Identification and Listing of Hazardous Waste." In this section, it will be noted, a waste is hazardous if it is (1) a listed waste or (2) a characteristic waste.

Listed Wastes

The regulations include four lists identified by the letters F, K, P, and U that together describe about 400 hazardous wastes. The F-list includes hazardous wastes from non-specific sources. Table I provides examples from the F-list along with their EPA hazardous waste numbers and hazard rating (T = toxic, I = ignitable). The K-list covers hazardous wastes from specific sources. Table II lists examples. The P-list includes acutely hazardous wastes. Table III lists a number of these. Note that some are familiar items in a chemistry laboratory. The U-list gives commercial chemical products, chemical intermediates, and off-specification chemical products (see Table IV). Many of these wastes are also found in academic laboratories.

Table II. Some Examples from the K-List: Hazardous Wastes from Specific Sources

K008	oven residue from production of chrome oxide green pigments
K013	bottom stream from acetonitrile column in production of acrylonitrile
K020	heavy ends from distillation of vinyl chloride in vinyl chloride monomer production
K032	wastewater treatment sludge from chlordane production

Note: See Chapter 1, Appendix II for the entire list.

Table III. Some Examples from the P-List: Acutely Hazardous Wastes

EPA ID Number	Compound
P022	carbon disulfide
P037	dieldrin
P042	epinephrine
P056	fluorine
P076	nitric oxide
P077	p. nitroaniline
P087	osmium tetroxide
P110	tetra ethyl lead

Note: See Chapter 1, Appendix II for the entire list.

Characteristic Wastes

What if a waste cannot be found on one of the RCRA lists? Maybe it is an unknown. To establish the nature of such a waste, it must be tested to determine if it has certain properties or characteristics that render it hazardous. The waste is hazardous if it has one or more of the characteristics listed in Table V. (Further details of the EP Toxicity contaminants are given in Table VI.)

If a waste cannot be identified according to these criteria, a sample should be submitted to a reputable testing laboratory for identification. Help might also be obtained from local or state EPA offices.

Table IV. The U-List: Some Examples of Commercial Chemical Products, Chemical Intermediates, and Off-Specification Chemical Products

EPA ID Number	Compound
U001	acetaldehyde[a] (common name)
U012	aniline
U019	benzene
U188	benzene, hydroxy
U044	chloroform
U001	ethanal[a] (IUPAC name)
U122	formaldehyde
U151	mercury

Note: See Chapter 1, Appendix II for the entire list.
[a]Note the cross-listing using both common and International Union of Pure and Applied Chemistry (IUPAC) names.

Table V. Characteristics of Hazardous Waste

Hazardous Characteristic	Waste Number	Brief Explanation
Ignitability	D001	Easily combustible or flammable liquid, flash point <60°C (<140°F). Solids that burn easily. (Solvents, paints, metal dust)
Corrosivity	D002	Dissolves metals or other materials or burns the skin, pH < 2 or > 12.5. (Acids and bases)
Reactivity	D003	Unstable or undergoes rapid or violent chemical reaction with water or other materials, releases toxic gases. (Explosives, sulfides, cyanides, picric acid, sodium metal)
EP Toxicity[a]	D004-D017	Extract contains high concentration of heavy metals and/or specific pesticides that could be released into ground water.

[a]See Table VI.

HOW TO DETERMINE GENERATOR STATUS

Once it has been determined that the waste produced by a facility is hazardous, the facility is, by definition, a hazardous waste generator.

To determine the status of a facility, and hence know which regulations apply, *the total hazardous waste the facility generates per calendar month must be measured*. Figure 1 shows the steps involved. In most cases, universities will produce 100–1000 kg of waste per month and will be classed as small quantity generators.

If a facility generates less than 100 kg of hazardous waste per month, it is not a small quantity generator. It may store up to 1000 kg on its property at

Table VI. Maximum Concentration of Contaminants for Characteristic of EP Toxicity

EPA hazardous waste number	Contaminant	Maximum concentration (mg/L)
D004	Arsenic	5.0
D005	Barium	100.0
D006	Cadmium	1.0
D007	Chromium	5.0
D008	Lead	5.0
D009	Mercury	0.2
D010	Selenium	1.0
D011	Silver	5.0
D012	Endrin	0.02
D013	Lindane	0.4
D014	Methoxychlor	10.0
D015	Toxaphene	0.5
D016	2,4-D	10.0
D017	2.4,5-TP Silvex	1.0

Source: 40 CFR 261.24.

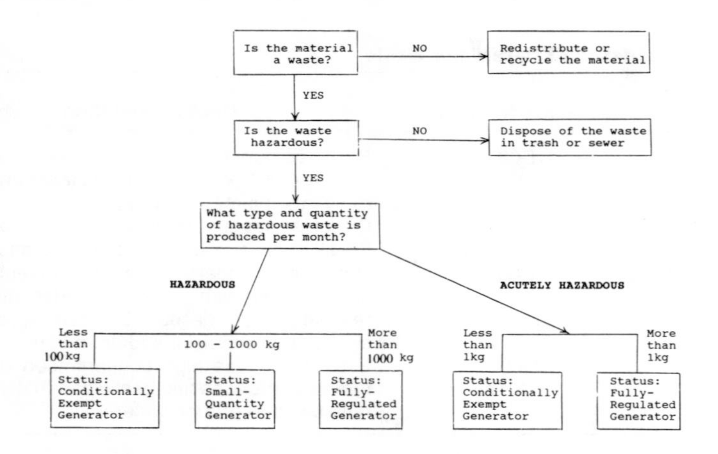

Figure 1. How to determine generator status. [Adapted with permission from "Hazardous Waste Management at Educational Institutions," Figure 2-2 copyright 1987 National Association of College and University Business Officers.]

any time. However, if more than 1000 kg are stored, it becomes a small quantity generator.

If a facility is a small quantity generator, it can accumulate up to 6000 kg in any 180 day period (270 day period if the waste is to be transported more than 200 miles). Otherwise, it needs a special storage permit.

HOW TO GET AN EPA IDENTIFICATION NUMBER

If a facility is a hazardous waste generator, it will eventually need the services of a transporter and the use of a Transfer, Storage, and Disposal Facility (TSDF).

To use such services, a facility must have an EPA ID number, obtained by requesting a "Notification of Hazardous Waste Activity" form from the EPA or a state agency (see Appendix III). The form must be completed, signed, and returned to the designated EPA office. A unique 12 character EPA ID number will then be assigned for use on all documents when shipping hazardous waste. If there are several waste generation sites at a facility, each may be required to have its own EPA ID number.

HOW IS HAZARDOUS WASTE MANAGED ON-SITE?

In brief, regulations for storage time, quantity of waste, and handling requirements must be followed.

Permits may need to be obtained for on-site storage, treatment, and/or disposal of the waste. It will also be necessary to ensure that adequate precau-

COURSE: ____________________ ROOM: ____________________
EXPT. #: ____________________ DATE: ____________________
PLEASE PLACE ONLY THE WASTE SPECIFIED ON THIS FORM IN THIS CONTAINER.

CONTENTS:

Hazardous Waste Management
692-3556
Southern Illinois University at Edwardsville
Edwardsville, Illinois 62026-1652

School of Sciences

Figure 2. Typical label design for waste collection container.

tions are taken to prevent accidents and cope with accidents should they occur.

Basic Waste Collection Procedure

- Use labeled and dated containers in the laboratories. A sample label is shown in Figure 2.
- Keep containers closed except to fill or empty.
- Keep an operational log book to record activities. A sample page is given in Figure 3.

Hazardous Waste Storage

A small quantity generator can store up to 6000 kg on-site in any 180 day period. A storage permit is needed if this amount or time is exceeded. A typical storage procedure would include the following:

- 55-gal drums in good condition
- "Hazardous Waste" label on containers with start-date
- containers kept closed except to fill or empty
- inspection of containers for leaks every week (see sample log, Figure 4)
- daily inspection of safety equipment (see sample log, Figure 5)
- written plans developed for emergencies

Date	Source	Contents	Waste Char.	Code	Sign
6-3-88	SL 2213	Silver nitrate solution, 500 mL	pH = 5	3355	JG
6-5-88	Alestle	Photo developer, 10 liters	neutral	3356	PH
6/7/88	Solvent Rm.	White powder - unknown	500 g	3357	PH
6/7/88	Res. Lab	Used Pump Oil	2¼ gal	3358	PH
6-10-88	Research	o-Tolidine reagent, 500 mL	pH = 1	3359	PH

Figure 3. Sample page from operational log.

Hazardous Waste Treatment

A special permit is unnecessary if accumulated waste is treated in 180 days (270 days), and container and safety regulations are complied with.

If these requirements are not met, a hazardous waste treatment permit must be obtained.

HAZARDOUS WASTE DISPOSAL

Hazardous waste may not be disposed of on-site unless a disposal permit has been obtained. There are circumstances in which waste can be disposed directly into sewer drains without a permit, but this is discouraged because it is not good management practice. Local wastewater or sewage treatment plants may be contacted for specific details.

Obtaining a permit to store, treat, or dispose of hazardous waste on-site (40 CFR Part 270) can be costly and time-consuming. Assistance in making the determination may be obtained by contacting regional EPA offices.

ACCIDENT PREVENTION AND ACCIDENT HANDLING

If hazardous waste is stored, steps must be taken to avoid fire, explosions, and release.

WEEKLY DRUM INSPECTION LOG

Date	Inspector	Condition of Containers	All Containers Closed	Signs of Leakage (incl. floors, drums, etc.)	Remarks (if corrective action is required	List Nature Description of Actions taken & Date taken
9/2/88	PH/RG	good	yes	none	Dirty floors	floors swept 9/4/88
9/11/88	PH/JW	good	yes	slight leak from waste solvent drum	vermiculite application to contain spill	vermiculite scraped up - 9/12/88
9/18/88	RG/JW	good	yes	none		5 drums taken for disposal off-site

Figure 4. Sample log.

Daily Safety Equipment and Facility Inspection Log

Date	Inspector	Item	Date	Date Check	Date when	Date Inspected.	Date	Remarks
		Goggles						
		Respirators						
		Safety Gloves						
		Zetec Gloves						
		Lab Coat or Overalls						
		Apron						
		Fire Extinguishers in place						
		Fire Blanket						
		Acid Neutralizer						
		Base Neutralizer						
		Sand w/ 10% Soda Ash						
		Plastic Bags						
		Spill Cart Supplies						
		Emergency Power Equipment						
		Main Power Unit						
		Hood Exhaust Switch						
		Shower						
		Eyewash Fountain						
		Face Shields						
		Body Shields						
		Scott Lab-Pak						
		Storage Containers						
		Distillation Appar.						
		* Room Order						

* Report excessive clutter in the Waste Management Area.

Figure 5. Sample log.

It is also necessary to have an emergency communication system, the appropriate fire equipment, and plans with local police and emergency response teams, etc.

Contingency plans must be drawn up. They must include emergency procedures, emergency phone numbers, name of emergency coordinator, and containment measures available.

A personnel training program must be in place along with a training manual and records of personnel training.

HOW TO DISPOSE OF HAZARDOUS WASTE OFF-SITE

The hauler and designated facility (TSDF) selected must have EPA ID numbers. Either the waste generator or the hauler must package and label wastes for shipping, and prepare a Hazardous Waste Manifest. Key elements of off-site hazardous waste disposal follow:

- *Hauler/TSDF* — The hauler and designated facility must be chosen carefully and their EPA ID numbers checked. The waste is still the responsibility of the waste generator.
- *Package* — Containers must be acceptable for transport and properly labeled.
- *Manifest* — This multiple copy document accompanies the waste, tracks waste from "cradle-to-grave," is signed by generator, hauler, and designated facility and is kept on file for three years.
- *Costs* — Costs can be alarming and continue to rise.

1977	$20/drum
1982	$175/drum
1984	$340/drum
1988	$800 – for a drum not correctly packaged!

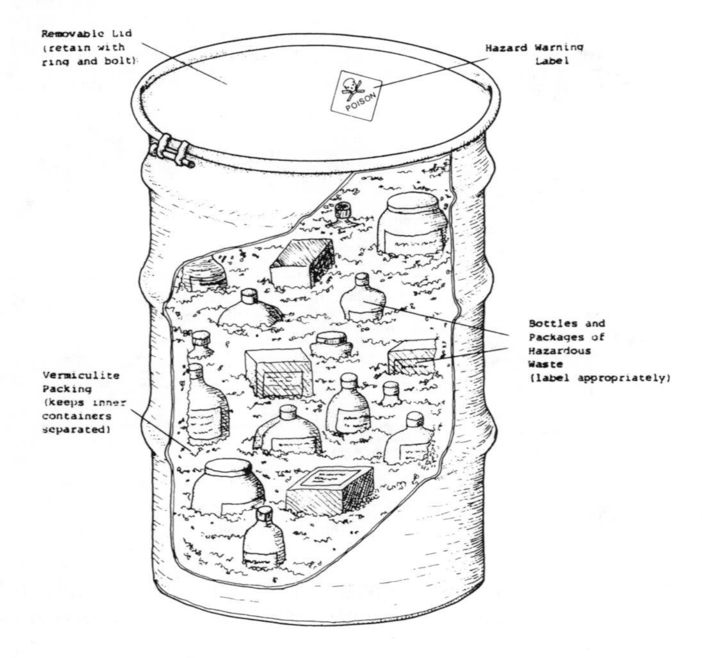

Figure 6. "Lab pack" in DOT 17C steel drum.

It is important to choose a reputable hauler and waste management facility, since even after waste has been removed, the waste generator is responsible for its proper management. The waste management facility is the final destination of hazardous waste for treatment, storage, or disposal.

Packing Hazardous Wastes for Shipment

The hauler can usually assist with the packing, but it will raise the cost of disposal. DOT regulations at 49 CFR Part 172 may also be referred to. There are two points to note:

1. Containers must be acceptable for transportation.
   - open-top 55-gal drum for lab packs (Figure 6)
   - closed-top 55-gal drum for solvents
2. Containers must be properly labeled.

The Uniform Hazardous Waste Manifest

The Manifest is a multicopy shipping document that must accompany hazardous waste shipments (Figure 7). It is designed so hazardous waste can

STATE OF ILLINOIS

ENVIRONMENTAL PROTECTION AGENCY DIVISION OF LAND POLLUTION CONTROL

P.O. BOX 19276 SPRINGFIELD, ILLINOIS 62794-9276 (217) 782-6761

State Form LPC 62 8/81 IL532-0610

FOR SHIPMENT OF HAZARDOUS, INFECTIOUS AND SPECIAL WASTE

PLEASE TYPE (Form designed for use on elite (12-pitch) typewriter) **EPA Form 8700-22 (Rev. 9-86)** Form Approved OMB No. 2050-0039 Expires 9-30-91

UNIFORM HAZARDOUS WASTE MANIFEST

1. Generator's US EPA ID No. ILD001234567 Manifest Document No. 00001
2. Page 1 of 1

Information in the shaded areas is not required by Federal law, but is required by Illinois law.

3. Generator's Name and Mailing Address — Location If Different:
ABC Generator Company
#1 Industrial Drive Your town, IL 62626

4. Generator's Phone (312) 444-4545

A. Illinois Manifest Document Number IL 4105561 MANIFEST FEE PAID

B. Illinois Generator's ID 0310000000

5. Transporter 1 Company Name: Town Hauler Co.
6. US EPA ID Number: ILD001200123

C. Illinois Transporter's ID 0010

D. (312) 555-4555 Transporter's Phone

7. Transporter 2 Company Name: Fast Haul, Inc.
8. US EPA ID Number: ILD002300234

E. Illinois Transporter's ID 0009

F. (312) 444-3444 Transporter's Phone

9. Designated Facility Name and Site Address:
A & B Landfill, Inc.
County Hwy #1 Your town, IL
10. US EPA ID Number: ILD002211331

G. Illinois Facility's ID 0310000102

H. Facility's Phone (312) 333-2333

11. US DOT Description (*Including Proper Shipping Name, Hazard Class, and ID Number*)	12. Containers No.	Type	13. Total Quantity	14. Unit Wt/Vol	I. Waste No.
a. Asbestos, ORMC - Not Hazardous	003	CM	00003	2	EPA HW Number XX; Authorization Number 846024
b. Waste Trichloroethylene ORM-A UN1710	005	DM	00110	1	EPA HW Number XXU228; Authorization Number 999997
c. Waste Sulphuric Acid Corrosinve Material UN1830	008	DM	00200	1	EPA HW Number XXU103; Authorization Number 854000
d.					EPA HW Number XX; Authorization Number

J. Additional Descriptions for Materials Listed Above
a - asbestos debris
b - trichloroethylen, off specification
c - sulphuric acid from catalyst regeneration

K. Handling Codes for Wastes Listed Above In Item # 14
1 = Gallons 2 = Cubic Yards
In Item 14

15. Special Handling Instructions and Additional Information

If waste listed in Item 11 (a through c) is undeliverable for any reason - return to generator.

16. **GENERATOR'S CERTIFICATION:** I hereby declare that the contents of this consignment are fully and accurately described above by proper shipping name and are classified, packed, marked, and labeled, and are in all respects in proper condition for transport by highway according to applicable international and national government regulations.

If I am a large quantity generator, I certify that I have a program in place to reduce the volume and toxicity of waste generated to the degree I have determined to be economically practicable and that I have selected the practicable method of treatment, storage, or disposal currently available to me which minimizes the present and future threat to human health and the environment; **OR**, if I am a small quantity generator, I have made a good faith effort to minimize my waste generation and select the best waste management method that is available to me and that I can afford.

Printed/Typed Name | Signature | Date Month Day Year

17. Transporter 1 Acknowledgement of Receipt of Materials
Printed/Typed Name | Signature | Date Month Day Year

18. Transporter 2 Acknowledgement of Receipt of Materials
Printed/Typed Name | Signature | Date Month Day Year

19. Discrepancy Indication Space

20. Facility Owner or Operator: Certification of receipt of hazardous materials covered by this manifest except as noted in item 19.
Printed/Typed Name | Signature | Date Month Day Year

GENERATOR

TRANSPORTER

FACILITY

In case of a spill call the Illinois Office of Emergency Response at 217/782-3637 and the National Response Center at 800/424-8802 or 202/426-[illegible]

This Agency is authorized to require, pursuant to Illinois Revised Statutes Chapter 111 [illegible] Section [illegible] that this information be submitted to the Agency. Failure to provide the information may result in [illegible] penalty against the owner or operator of not to exceed $25,000 per day of violation. Falsification of this information may result in a fine up to $[illegible] per day of violation and imprisonment up to [illegible] years. [illegible] Forms Management Center.

COPY 1. TSD MAIL TO GENERATOR

Figure 7. The Manifest.

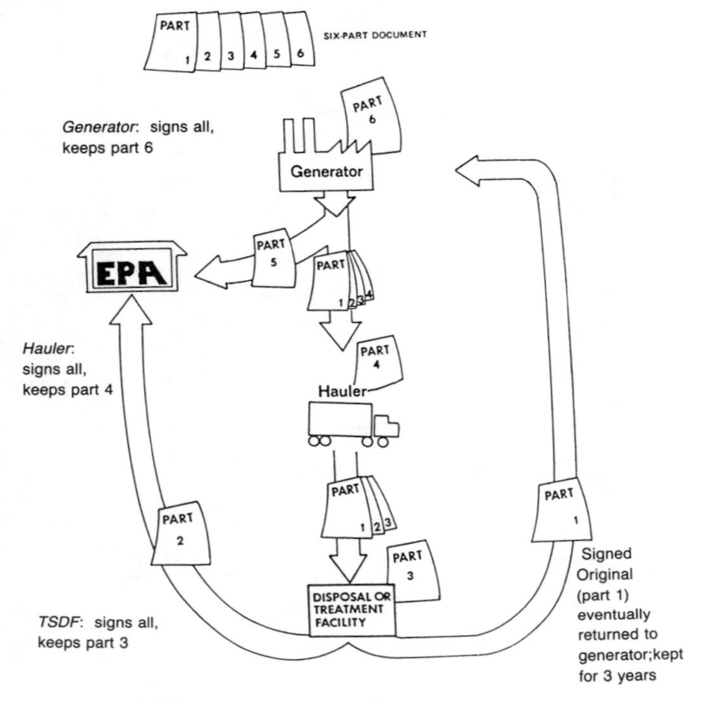

Figure 8. A typical hazardous waste manifest system. [Modification of an illustration from the Illinois Environmental Protection Agency. Used with permission.]

be tracked from generator to final destination or disposal—"cradle-to-grave." Points to note (Figure 8):

- The generator, hauler, and operator from the designated facility must each sign the document.
- The designated facility operator must send a copy back to the waste generator as notification that the shipment has reached its destination. This copy must be kept on file for three years.
- If the signed copies from the designated facility are not received within 35 days, contact the facility or transporter. If the copies are not received in 45 days, a manifest discrepancy report must be filed with the EPA and/or appropriate state agency.

GOOD HOUSEKEEPING = SAFE ENVIRONMENT

Waste should be managed properly. Some helpful hints follow:

- A waste minimization program must be in effect.
- Keep hazardous/nonhazardous wastes separated and clearly labeled.
- Avoid leaks and spills—cleanup operations can be costly.

- Use the minimum amount of hazardous chemicals necessary to get the job done.
- Recycle or redistribute excess chemicals.
- Empty original containers—then discard.

It is necessary to cooperate with State and Federal Inspection Agencies—they can help identify and correct problems. These are some pointers that will facilitate waste handling:

- Document kinds and amounts of hazardous wastes generated.
- Document methods of determination.
- Include EPA ID number on all shipments of waste.
- Store wastes in proper containers.
- Label and date containers—keep them closed except to fill and empty.
- Keep signed copies of Uniform Hazardous Waste Manifests.
- Designate an emergency coordinator.
- Post emergency telephone numbers and response procedures.
- Provide evidence for a training program for employees.

BIBLIOGRAPHY

Allen, R. O. "Waste Disposal in the Laboratory: Teaching Responsibility and Safety," *J. Chem. Educ.* 60:A81 (1983).

Armour, M. A. "Chemical Waste Management and Disposal,"*J. Chem. Educ.* 65:A64 (1988).

Ashbrook, P., and P. Reinhardt. "Hazardous Wastes in Academia," *Envir. Sci. Technol.* 19:1150 (1985).

Chlad, F. L. "Hazardous Chemical Waste Disposal and the Impact of RCRA," *J. Chem. Educ.* 59:331 (1982).

Fischer, K. E. "Self-Audits of Hazardous Waste Operations in Laboratories," *J. Chem Educ.* 64:A207 (1987).

Gerlovich, J. A., and J. Miller. "Safe Disposal of Unwanted School Chemicals—A Proven Plan," *J. Chem. Educ.* 66:433 (1989).

Lowry G. G., and R. C. Lowry. *Lowry's Handbook of Right-to-Know and Emergency Planning* (Chelsea, MI: Lewis Publishers, Inc., 1988).

McKusick, B. "Classification of Unlabelled Laboratory Waste for Disposal," *J. Chem. Educ.* 63:A128 (1986).

________. "Procedures for Laboratory Destruction of Chemicals," *J. Chem. Educ.* 61:A152 (1984).

"RCRA and Laboratories" and "Less Is Better" (pamphlets) and other related literature (The American Chemical Society, Office of Federal Regulatory Programs, 1155 16th Street NW, Washington, DC 20036).

Sanders, H. T. "Hazardous Wastes in Academic Labs," *C & E News*, Special Report (3 February 1986), pp.21–31.

"Understanding the Small-Quantity Generator Hazardous Waste Rules: A Handbook for Small Business," Office of Solid Waste and Emergency Response, U.S. EPA/530-SW-86–019 (1986).

Young, J. A. "Academic Laboratory Waste Disposal. Yes, You Can Get Rid of That Stuff Legally!," *J. Chem. Educ.* 60:490, (1983).

CHAPTER 1, APPENDIX 1

Useful Canadian Addresses

NATIONAL ADDRESSES

AGRICULTURE CANADA	Animal Health Division Chief of Veterinary Biology 801 Fallowfield Road, Box 11300 Station "H" Nepean ON K2H 7T9
	Safety, Health and Security and Fire Prevention Section Sir John Carling Building Ottawa ON K1A 0C5 (613) 993-6524 FACS: (613) 998-1312 For pesticide information: 1-800-267-6315
ATOMIC ENERGY CONTROL BOARD	Office of Public Information P.O. Box 1046 Ottawa ON K1P 5S9
CANADIAN CENTRE FOR OCCUPATIONAL HEALTH AND SAFETY	250 Main Street East (at Wellington) Hamilton ON L8N 1H6 Toll free number: 1-800-263-8276 or dial (416) 572-2981 Hours: Monday to Friday, 8:30 a.m. to 5:00 p.m. Telex: 061-8532 CCINFO, the Centre's national computerized information retrieval service.

ENVIRONMENT CANADA Atlantic Region	Conservation and Protection (Waste Management) 5th Floor, Queen Square 45 Alderney Drive Dartmouth NS B2Y 2N6 (902) 426-3593
Newfoundland District Office	P.O. Box 5037 St. John's NF A1C 5V3 (709) 772-5488
New Brunswick District Office	364 Argyle Street Fredericton NB E3B 1T9 (506) 452-3286
Prince Edward Island District Office	P.O. Box 1115, Dominion Building Charlottetown PE C1A 7M8 (902) 566-7042
Nova Scotia District Office	5th Floor, Queen Square 45 Alderney Drive Dartmouth NS B2Y 2N6 (902) 426-6086
Quebec Region	1179 de Bleury St. Montreal QC H3B 3H9 (514) 283-0178
Ontario Region	25 St. Clair Avenue East 7th Floor Toronto ON M4T 1M2 (416) 973-1055
Ottawa District Office	River Road Environmental Technology Centre 3439 River Road Ottawa ON K1A 1C8 (613) 991-1954
Western and Northern Region	Twin Astria No.2, Second Floor 4999 – 98 Avenue Edmonton AB T6B 2X3 (403) 468-8041

Alberta District Office	Twin Astria No.2, Second Floor 4999 - 98 Avenue Edmonton AB T6B 2X3 (403) 468-8007
Saskatchewan District Office	2nd Floor, 1901 Victoria Avenue Region SK S4P 3R4 (306) 780-6464
Manitoba District Office	800 Kensington Building 275 Portage Avenue Winnipeg MB R3B 2B3 (204) 949-4811
Northwest Territories District Office	9th Floor, Bellanca Building Box 370 Yellowknife NWT X1A 2N3 (403) 873-3456
Inuvik Sub-Office	211, Federal Building P.O. Box 1086 Inuvik NWT X0E 0T0 (403) 979-2313
Frobisher Bay Sub-Office	No. 2, Dublanco Building Box 384 Frobisher Bay NWT X0A 0H0 (403) 979-6349
Pacific Region	3rd Floor, Kapilano 100-Park Royal West Vancouver BC V7T 1A2 (604) 666-0064
Yukon Branch	Room 225 - Federal Building Whitehorse YK Y1A 2B5 (403) 667-6487

HEALTH AND WELFARE CANADA — Regional Directors - Medical Services

National Capital Region	(Ottawa)	(613) 990-0622
Atlantic Region	(Halifax)	(902) 426-6141
Quebec Region	(Montreal)	(514) 283-6418
Ontario Region	(Ottawa)	(416) 973-1073
Manitoba Region	(Winnipeg)	(203) 949-4172
Saskatchewan Region	(Regina)	(306) 780-5413
Alberta Region	(Edmonton)	(403) 420-2690

Pacific Region	(Vancouver)	(604) 666-3235
NWT Region	(Yellowknife)	(403) 873-7025
Yukon Region	(Whitehorse)	(403) 668-6461

LABOUR CANADA	Occupational Health and Safety Branch Ottawa ON K1A 0J2 (819) 997-3520
MEDICAL RESEARCH COUNCIL OF CANADA	Jeanne Mance Bldg. Ottawa, ON K1A 0W9 (613) 954-1812
STATISTICS CANADA – LABOUR DIVISION	National Work Injuries Statistics Program Ottawa ON K1A 0T6 (613) 990-9900
TRANSPORT CANADA	CANUTEC (Canadian Transport Emergency Centre) Transport Dangerous Goods Place de Ville Ottawa ON K1A 0N5 Telex: 053-3130 (DOT-OTT) Emergency phone number: (613) 996-6666 (collect) Other information: (613) 992-4624

Headquarters	(Ottawa)	(613) 998-6540
Maritime Regional Office	(Halifax)	(902) 426-9351
Eastern Regional Office	(Montreal)	(514) 283-0696
Central Regional Office	(Toronto)	(416) 973-4599
Prairies Regional Office	(Winnipeg)	(204) 949-8839
Western Regional Office	(Saskatoon)	(306) 975-5059
Pacific Regional Office	(New Westminster)	(604) 666-2955

PROVINCIAL ADDRESSES

NOVA SCOTIA

LABOUR CANADA	Suite 300 6009 Quinpool Road P.O. Box 8628 Halifax NS B3K 5M3 (902) 426-3833

	Government of Canada Building 308 George Street Sydney NS B1P 1J8 (902) 564-7130
TRANSPORTATION OF DANGEROUS GOODS	Motor Vehicle Inspection Division Road Transport Inspection Section Department of Transportation 6061 Young Street P.O. Box 156 Halifax NS B3J 2M4
OCCUPATIONAL HEALTH AND SAFETY DIVISION	Department of Labour P.O. Box 697 Halifax NS B3J 2T8 (902) 424-4328 Occupational Health (902) 424-4428 Training (902) 424-2726 Occupational Safety (902) 424-7649
POISON CONTROL CENTRE	The Izaak Walton Killam Hospital for Children 5850 University Avenue Halifax NS B3J 3G9 (902) 428-8161

NEW BRUNSWICK

LABOUR CANADA	Professional Arts Building 100 Arden Street P.O. Box 2967, Station A Moncton NB E1C 8T8 (506) 857-6640
	Professional Arts Building 5th Floor 100 Arden Street Moncton NB E1C 4B7 (506) 857-7424

TRANSPORTATION OF DANGEROUS GOODS

Deputy Registrar of Motor Vehicles
Motor Vehicle Division
Department of Transportation
Kings Place, York Tower
York Street
P.O. Box 6000
Fredericton NB E3B 5H1

OCCUPATIONAL HEALTH AND SAFETY DIVISION

Department of Labour and Human Resources
P.O. Box 6000
Fredericton NB E3B 5H1
(506) 453-2467

- Policy, Planning and Administration Division (506) 453-2467
- Health and Safety Services Division (506) 453-2467

POISON CONTROL CENTRE

Saint John Regional Hospital
Waterloo Street
Box 2100
Saint John NB E2L 4L2
(506) 648-7111

PRINCE EDWARD ISLAND

TRANSPORTATION OF DANGEROUS GOODS

Director
Highway Safety Division
Transportation and Public Works
17 Haviland Street
P.O. Box 2000
Charlottetown PE C1A 7N8

OCCUPATIONAL HEALTH AND SAFETY DIVISION

Department of Fisheries and Labour
P.O. Box 2000
Charlottetown PE C1A 7N8
(902) 892-3493

POISON CONTROL CENTRE

Queen Elizabeth Hospital
P.O. Box 6600
Charlottetown PE C1A 8T5
(902) 566-6250

NEWFOUNDLAND

LABOUR CANADA	Sir Humphrey Gilbert Building P.O. Box 5278 St.John's NF A1C 5W1 (709) 722-5022
TRANSPORTATION OF DANGEROUS GOODS	Highway Equipment Manager Department of Transportation P.O. Box 4750 St.John's NF A1C 5T7
OCCUPATIONAL HEALTH AND SAFETY DIVISION	Department of Labour P.O. Box 4750 St.John's NF A1C 5T7 (709) 576-2721 • Occupational Health and Safety Services Branch (709) 576-2694
POISON CONTROL CENTRE	The Dr. Charles A. Janeway Child Health Centre Newfoundland Drive St.John's NF A1A 1R8 (709) 722-1110

QUEBEC

LABOUR CANADA	Guy-Favreau Complex Suite 101, West Tower 200 Dorchester Blvd. West Montreal QC H2Z 1X4 (514) 283-8538
	390 South Dorchester Street P.O. Box 3279 Saint Roch Station Quebec QC G1K 6Y7 (418) 648-7707
TRANSPORTATION OF DANGEROUS GOODS	Direction du transport routier des marchandises Ministère des Transports 700 St-Cyrille Est 22e étage Quebec QC G1R 5H1

COMMISSION DE LA SANTE ET LA SECURITE DU TRAVAIL (CSST)	1199 de Bleury C.P. 6056, Succ. A Montreal QC H3C 4E1 (514) 873-3503 • Toxicological Index (514) 873-6374 • Prevention-Inspection Division (514) 873-5211 • Communications Division (514) 873-7345
ENVIRONNEMENT – QUEBEC	5199 rue Sherbrooke Est Bureau 3860 Montréal QC H1T 3X9 (514) 873-3636 URGENCE: (514) 873-3454 (514) 643-4595
POISON CONTROL CENTRE	1-800-463-5060 (toll free)

ONTARIO

LABOUR CANADA	Co-operative Building Suite 325 4211 Yonge Street Willowdale ON M2P 2A9 (416) 224-3850
	Toronto West District Office Suite 708 6711 Mississauga Road Mississauga ON L5N 2N3 (416) 858-8615
	Toronto East District Office 8th Floor 200 Town Centre Court Scarborough ON M1P 4X8 (416) 973-4500

SBI Bldg., 11th Floor
Billings Bridge Plaza
2323 Riverside Drive
Ottawa ON K1H 8L5
(613) 998-9083

Dominion Public Bldg.
Room 505
457 Richmond Street
London ON N6A 3E3
(519) 679-4047

Room 337
33 Court Street South
Thunder Bay ON P7B 2W6
(807) 345-5474

TRANSPORTATION OF DANGEROUS GOODS

Coordinator, Dangerous Goods Project
Ontario Ministry of Transportation & Communications
Room 212
West Building
1201 Wilson Avenue
Downsview ON M3M 1J8

OCCUPATIONAL HEALTH AND SAFETY DIVISION

Ministry of Labour
400 University Ave.
Toronto ON M7A 1T7
(416) 965-9450

- Occupational Health Branch (416) 965-3211
- Construction Health and Safety Branch (416) 965-7161
- Industrial Health and Safety Branch (416) 965-4125
- Special Studies and Services Branch (416) 965-2493

POISON CONTROL CENTRES

Hospital for Sick Children
555 University Avenue
Toronto ON M5G 1X8
(416) 979-1900

Children's Hospital of Eastern Ontario
401 Smyth Road
Ottawa ON K1H 8L1
(613) 521-4040

MANITOBA

LABOUR CANADA

Room 400
303 Main Street
Winnipeg MB R3C 3G7
(204) 949-7223

Room 304
1 Wesley Avenue
Winnipeg MB R3C 4C6
(204) 949-2409

TRANSPORTATION OF DANGEROUS GOODS

Dangerous Goods Handling and Transportation Information
Environmental Management Division
Department of Environment
Workplace Safety & Health
Building 2
139 Tuxedo Avenue, Box 7
Winnipeg MB R3N 0H6

WORKPLACE SAFETY AND HEALTH DIVISION

Department of Environment and Workplace Safety and Health
960 - 330 St. Mary Ave.
Winnipeg MB R3C 3Z5
(204) 945-3605

- Safety and Health Inspectorate
  (204) 945-3602
- Industrial Hygiene Branch
  (204) 945-5765
- Educational Services Branch
  (204) 945-3613

POISON CONTROL CENTRES

Provincial Poison Information Centre
(204) 787-2591

Health Sciences Children's Centre
685 Bannatyne Avenue
Winnipeg MB R3E 0W1
(204) 787-2444

SASKATCHEWAN

LABOUR CANADA	Financial Building Room 301 2101 Scarth Street Regina SK S4P 2H9 (306) 780-5408
	Federal Building Room 306 1st Avenue - 22nd Street Saskatoon SK S7K 0E1 (306) 975-4303
TRANSPORTATION OF DANGEROUS GOODS	Dangerous Goods Officer Planning Support Branch Transportation Planning and Research Division Saskatchewan Highways and Transportation 1855 Victoria Avenue Regina SK S4P 3V5
OCCUPATIONAL HEALTH AND SAFETY BRANCH	Department of Labour Saskatchewan Place 1870 Albert St. Regina SK S4P 3V7 (306) 787-4481
POISON CONTROL CENTRES	Regina General Hospital Poison Control Centre 1440 - 14th Avenue Regina SK S4P 0W5 (306) 359-4545
	Saskatoon University Hospital Poison Control Centre Saskatoon SK S7N 0W8 (306) 343-3323

ALBERTA

LABOUR CANADA	Energy Square 3rd Floor 10109 - 106 Street Edmonton AB T5J 3L7 (403) 420-2989 Room 574 220 - 4th Avenue S.E. P.O. Box 2800 Station M Calgary AB T2P 3C2 (403) 292-4566
ALBERTA PUBLIC SAFETY EMPLOYEES	Compliance Information Centre (for information about classification, documentation, labeling and placards, dangerous goods routes, other regulatory requirements) • Edmonton (403) 422-9600 • Anywhere else in Alberta 1-800-272-9600
TRANSPORTATION OF DANGEROUS GOODS	Dangerous Goods Control Alberta Disaster Services 144, 14315 - 118 Avenue Edmonton AB T5L 2M3
OCCUPATIONAL HEALTH AND SAFETY DIVISION	Ministry of Community and Occupational Health 10709 Jasper Ave. Edmonton AB T5J 3N3 (403) 427-6971 • Work Site Services (403) 427-5566 • Occupational Health Services (403) 427-8067 • Research and Education Services (403) 427-5549
POISON CONTROL CENTRES	Royal Alexandra Hospital 10240 Kingsway Avenue Edmonton AB T5H 3V9 (403) 474-3431

	University of Alberta Hospital 112nd Street and 84th Avenue Edmonton AB T6G 2B7 (403) 432-8410
	Calgary General Hospital 841 Centre Avenue East Calgary AB T2E 0A1 (403) 262-5982
	Emergency Department Foothills General Hospital 1403 – 29th Street, N.W. Calgary AB T2N 2T9 (403) 270-1315

BRITISH COLUMBIA

LABOUR CANADA	7th Floor 750 Cambie Street Vancouver BC V6B 2P2 (604) 666-0656
TRANSPORTATION OF DANGEROUS GOODS	Director, Administration and Road Safety Motor Vehicle Department Ministry of Transportation and Highways 2631 Douglas Street Victoria BC V8T 5A3
OCCUPATIONAL HEALTH AND SAFETY DIVISION	(Workers' Compensation Board, Department of Labour) 6951 Westminster Highway Richmond BC V7C 1C6 (604) 273-2266
POISON CONTROL CENTRES	B.C. Drug and Poison Information Centre St.Paul's Hospital 1081 Burrard Street Vancouver BC (604) 682-5050

	Emergency Department Royal Jubilee Hospital Victoria BC (604) 595-9212

NORTHWEST TERRITORIES

TRANSPORTATION OF DANGEROUS GOODS	Pollution Control Engineer Pollution Control Division Department of Renewable Resources Government of the Northwest Territories Yellowknife NWT X1A 2L9
SAFETY DIVISION	(Department of Justice and Public Services) P.O. Box 1320 Yellowknife NWT X1A 2L9 (403) 873-7619
POISON CONTROL CENTRE	Stanton Yellowknife Hospital Yellowknife NWT X0E 1H0 (403) 920-4111

YUKON

TRANSPORTATION OF DANGEROUS GOODS	Administrator, Transport Services Department of Community and Transportation Services P.O. Box 2703 Whitehorse YK Y1A 2C6
OCCUPATIONAL HEALTH AND SAFETY DIVISION	(Department of Justice) P.O. Box 2703 Whitehorse YK Y1A 2C6 (403) 667-5811
POISON CONTROL CENTRE	Whitehorse General Hospital Emergency Department Whitehorse YK (403) 668-9444

INTERNATIONAL ADDRESSES

CENTRES FOR DISEASE CONTROL (CDC)
Clefpon Rd.
Atlanta, GA
30333

INTERNATIONAL AGENCY FOR RESEARCH ON CANCER
150, Cours Albert Thomas
69372 Lyon Cedex 08
France

NATIONAL INSTITUTE OF HEALTH (NIH)
Bethesda, MD
20025

WORLD HEALTH ORGANIZATION (WHO)
525 - 23rd St. N.W.
Washington, DC
20037

TRANSPORTATION OF HAZARDOUS WASTES

FEDERAL AGENCIES

(For International Movement of Hazardous Wastes)

NATIONAL HEADQUARTERS

Environment Canada
Waste Management Division
351 St. Joseph Blvd., 13th floor
Hull, PQ K1A 1C8
(819) 997-3378

Transport Canada
Transport Dangerous Goods Directorate
Tower B, Place de Ville
Ottawa, Ontario K1A 0N5
(613) 992-4624

Environment Canada Regional Offices
Bureaux Régionaux d'Environnement Canada

Atlantic Region
Contaminants and Hazardous Waste
Environment Canada
5th floor, Queen's Square
45 Alderney Drive
Dartmouth (Nova Scotia)
B2Y 2N6
(902) 426-6141

Région du Québec
Urgences et résidues
Environnement Canada
1179 Bleury - 1iéme étage
Place des Arts
Montréal (Québec)
H3B 3H9
(514) 283-4684

Ontario Region
Environmental Contaminants
Environment Canada
Arthur Meighan Bldg. 7th floor
25 St. Clair Avenue East
Toronto (Ontario)
M4T 1M2
(416) 973-5840

Western and Northern Region
Waste Management
Environment Canada
9942 - 198 Street, 9th Floor
Edmonton (Alberta)
T5K 2J5
(403) 420-2591

Pacific and Yukon Region
Environmental Services
Environment Canada
3rd Floor, Kapilano 100
Park Royal
West Vancouver (British Columbia)
V7T 1A2
(604) 666-6711

PROVINCIAL REGULATORY AGENCIES

(For Inter & Intra-Provincial Movement of Hazardous Wastes)

Provincial Authorities – Autorités provinciales

Alberta
Waste Management Branch
Pollution Control Division
Alberta Environment
Oxbridge Place
9820 106 Street
Edmonton (Alberta)
T5K 2J6
(403) 427-5868

British Columbia
Colombie Britannique
Waste Management Branch
Ministry of the Environment
Parliament Building
Victoria (British Columbia)
V8V 1X5
(604) 387-9953

Manitoba
Department of Environment
Workplace Safety and Health
Environmental Management Division
Box 7 - Building 2
139 Tuxedo Avenue
Winnipeg (Manitoba)
R3N 0H6
(204) 945-7100

New Brunswick
Nouveau Brunswick
Municipal Affairs and Environment
Pollution Control Branch
P.O. Box 6000
Fredericton (New Brunswick)
E3B 5H1
(506) 453-2861

Newfoundland
Terre-Neuve
Department of Environment
Elizabeth Tower
Elizabeth Avenue
St. John's (Newfoundland)
A1C 5T7
(709) 576-2559

Northwest Territories
Territorires du Nord Quest
Pollution Control Division
Department of Renewable Resources
Box 1320
Yellowknife (N.W.T.)
X1A 2L9
(403) 873-7654

Nova Scotia
Nouvelle Ecosse
Department of Environment
P.O. Box 2107
Halifax (Nova Scotia)
B3J 3B7
(902) 424-5300

Ontario
Ontario Ministry of the Environment
Waste Management Branch
135 St. Clair Avenue West
Toronto (Ontario)
M4V 1P5
(416) 965-9668

Prince Edward Island
Ile du Prince Edouard
Department of Community and Cultural Affairs
P.O. Box 2000
Charlottetown (P.E.I.)
C1A 7N8
(902) 892-0311

Québec
Direction de la coordination et du contrôle
Ministère de l'environnement
3900 rue Marly
Boite postale 15
Ste-Foy (Québec)
G1X 4E4
(418) 643-7456

Saskatchewan
Land Protection Branch
Saskatchewan Environment
3085 Albert Street
Regina (Saskatchewan)
S4S 0B1
(306) 787-5811

Yukon
Minister of Community and Transportation Service
Government of the Yukon
P.O. Box 2703
Whitehorse (Yukon)
Y1A 2C6
(403) 667-3032

[Information in Appendix I provided by The Chemical Institute of Canada, Ottawa, 1987.]

CHAPTER 1, APPENDIX II

Listed Hazardous Wastes

Source: Reprinted from 40 CFR 261.31, 261.32, and 261.33.

SUBPART D—LISTS OF HAZARDOUS WASTES

§ 261.30 General.

(a) A solid waste is a hazardous waste if it is listed in this subpart, unless it has been excluded from this list under §§ 260.20 and 260.22.

(b) The Administrator will indicate his basis for listing the classes or types of wastes listed in this Subpart by employing one or more of the following Hazard Codes:

Ignitable Waste	(I)
Corrosive Waste	(C)
Reactive Waste	(R)
EP Toxic Waste	(E)
Acute Hazardous Waste	(H)
Toxic Waste	(T)

Appendix VII identifies the constituent which caused the Administrator to list the waste as an EP Toxic Waste (E) or Toxic Waste (T) in §§ 261.31 and 261.32.

(c) Each hazardous waste listed in this subpart is assigned an EPA

Hazardous Waste Number which precedes the name of the waste. This number must be used in complying with the notification requirements of Section 3010 of the Act and certain recordkeeping and reporting requirements under Parts 262 through 265 and Part 270 of this chapter.

(d) The following hazardous wastes listed in § 261.31 or § 261.32 are subject to the exclusion limits for acutely hazardous wastes established in § 261.5: EPA Hazardous Wastes Nos. FO20, FO21, FO22, FO23, FO26, and FO27.

[45 FR 33119, May 19, 1980, as amended at 48 FR 14294, Apr. 1, 1983; 50 FR 2000, Jan. 14, 1985]

§261.31 Hazardous wastes from non-specific sources.

The following solid wastes are listed hazardous wastes from non-specific sources unless they are excluded under §§ 260.20 and 260.22 and listed in Appendix IX.

Industry and EPA hazardous waste No.	Hazardous waste	Hazard code
Generic:		
F001	The following spent halogenated solvents used in degreasing: Tetrachloroethylene, trichloroethylene, methylene chloride, 1,1,1-trichloroethane, carbon tetrachloride, and chlorinated fluorocarbons; all spent solvent mixtures/blends used in degreasing containing, before use, a total of ten pe cent or more (by volume) of one or more of the above halogenated solvents or those solvents listed in F002, F004, and F005; and still bottoms from the recovery of these spent solvents and spent solvent mixtures.	(T)
F002	The following spent halogenated solvents: Tetrachloroethylene, methylene chloride, trichloroethylene, 1,1,1-trichloroethane, chlorobenzene, 1,1,2-trichloro-1,2,2-trifluoroethane, ortho-dichlorobenzene, trichlorofluoromethane, and 1,1,2-trichloroethane; all spent solvent mixtures/blends containing, before use, a total of ten percent or more (by volume) of one or more of the above halogenated solvents or those listed in F001, F004, or F005; and still bottoms from the recovery of these spent solvents and spent solvent mixtures.	(T)
F003	The following spent non-halogenated solvents: Xylene, acetone, ethyl acetate, ethyl benzene, ethyl ether, methyl isobutyl ketone, n-butyl alcohol, cyclohexanone, and methanol; all spent solvent mixtures/blends containing, before use, only the above spent non-halogenated solvents; and all spent solvent mixtures/blends containing, before use, one or more of the above non-halogenated solvents, and, a total of ten percent or more (by volume) of one or more of those solvents listed in F001, F002, F004, and F005; and still bottoms from the recovery of these spent solvents and spent solvent mixtures.	(I)*
F004	The following spent non-halogenated solvents: Cresols and cresylic acid, and nitrobenzene; all spent solvent mixtures/blends containing, before use, a total of ten percent or more (by volume) of one or more of the above non-halogenated solvents or those solvents listed in F001, F002, and F005; and still bottoms from the recovery of these spent solvents and spent solvent mixtures.	(T)
F005	The following spent non-halogenated solvents: Toluene, methyl ethyl ketone, carbon disulfide, isobutanol, pyridine, benzene, 2-ethoxyethanol, and 2-nitropropane; all spent solvent mixtures/blends containing, before use, a total of ten percent or more (by volume) of one or more of the above non-halogenated solvents or those solvents listed in F001, F002, or F004; and still bottoms from the recovery of these spent solvents and spent solvent mixtures.	(I,T)
F006	Wastewater treatment sludges from electroplating operations except from the following processes: (1) Sulfuric acid anodizing of aluminum; (2) tin plating on carbon steel; (3) zinc plating (segregated basis) on carbon steel; (4) aluminum or zinc-aluminum plating on carbon steel; (5) cleaning/stripping associated with tin, zinc and aluminum plating on carbon steel; and (6) chemical etching and milling of aluminum.	(T)

Industry and EPA hazardous waste No.	Hazardous waste	Hazard code
F019	Wastewater treatment sludges from the chemical conversion coating of aluminum	(T)
F007	Spent cyanide plating bath solutions from electroplating operations	(R, T)
F008	Plating bath residues from the bottom of plating baths from electroplating operations where cyanides are used in the process.	(R, T)
F009	Spent stripping and cleaning bath solutions from electroplating operations where cyanides are used in the process.	(R, T)
F010	Quenching bath residues from oil baths from metal heat treating operations where cyanides are used in the process.	(R, T)
F011	Spent cyanide solutions from salt bath pot cleaning from metal heat treating operations.	(R, T)
F012	Quenching waste water treatment sludges from metal heat treating operations where cyanides are used in the process.	(T)
F024	Wastes, including but not limited to, distillation residues, heavy ends, tars, and reactor clean-out wastes from the production of chlorinated aliphatic hydrocarbons, having carbon content from one to five, utilizing free radical catalyzed processes. [This listing does not include light ends, spent filters and filter aids, spent dessicants, wastewater, wastewater treatment sludges, spent catalysts, and wastes listed in § 261.32.].	(T)
F020	Wastes (except wastewater and spent carbon from hydrogen chloride purification) from the production or manufacturing use (as a reactant, chemical intermediate, or component in a formulating process) of tri- or tetrachlorophenol, or of intermediates used to produce their pesticide derivatives. (This listing does not include wastes from the production of Hexachlorophene from highly purified 2,4,5-trichlorophenol.).	(H)
F021	Wastes (except wastewater and spent carbon from hydrogen chloride purification) from the production or manufacturing use (as a reactant, chemical intermediate, or component in a formulating process) of pentachlorophenol, or of intermediates used to produce its derivatives.	(H)
F022	Wastes (except wastewater and spent carbon from hydrogen chloride purification) from the manufacturing use (as a reactant, chemical intermediate, or component in a formulating process) of tetra-, penta-, or hexachlorobenzenes under alkaline conditions.	(H)
F023	Wastes (except wastewater and spent carbon from hydrogen chloride purification) from the production of materials on equipment previously used for the production or manufacturing use (as a reactant, chemical intermediate, or component in a formulating process) of tri- and tetrachlorophenols. (This listing does not include wastes from equipment used only for the production or use of Hexachlorophene from highly purified 2,4,5-trichlorophenol.).	(H)
F026	Wastes (except wastewater and spent carbon from hydrogen chloride purification) from the production of materials on equipment previously used for the manufacturing use (as a reactant, chemical intermediate, or component in a formulating process) of tetra-, penta-, or hexachlorobenzene under alkaline conditions.	(H)
F027	Discarded unused formulations containing tri-, tetra-, or pentachlorophenol or discarded unused formulations containing compounds derived from these chlorophenols. (This listing does not include formulations containing Hexachlorophene sythesized from prepurified 2,4,5-trichlorophenol as the sole component.).	(H)
F028	Residues resulting from the incineration or thermal treatment of soil contaminated with EPA Hazardous Waste Nos. F020, F021, F022, F023, F026, and F027.	(T)

*(I,T) should be used to specify mixtures containing ignitable and toxic constituents.

[46 FR 4617, Jan. 16, 1981, as amended at 46 FR 27477, May 20, 1981; 49 FR 5312, Feb. 10, 1984; 49 FR 37070, Sept. 21, 1984; 50 FR 665, Jan. 4, 1985; 50 FR 2000, Jan. 14, 1985; 50 FR 53319, Dec. 31, 1985; 51 FR 2702, Jan. 21, 1986; 51 FR 6541, Feb. 25, 1986]

EFFECTIVE DATE NOTE: At 51 FR 6541, Feb. 25, 1986, in § 261.31, waste streams "F002" and "F005" in the subgroup "Generic" were revised, effective August 25, 1986. For the convenience of the user, the superseded text is set forth as follows:

§ 261.32 Hazardous wastes from specific sources.

The following solid wastes are listed hazardous wastes from specific sources unless they are excluded under §§ 260.20 and 260.22 and listed in Appendix IX.

Industry and EPA hazardous waste No.	Hazardous waste	Hazard code
Wood preservation: K001	Bottom sediment sludge from the treatment of wastewaters from wood preserving processes that use creosote and/or pentachlorophenol.	(T)
Inorganic pigments:		
K002	Wastewater treatment sludge from the production of chrome yellow and orange pigments.	(T)
K003	Wastewater treatment sludge from the production of molybdate orange pigments	(T)
K004	Wastewater treatment sludge from the production of zinc yellow pigments	(T)
K005	Wastewater treatment sludge from the production of chrome green pigments	(T)
K006	Wastewater treatment sludge from the production of chrome oxide green pigments (anhydrous and hydrated).	(T)

Industry and EPA hazardous waste No.	Hazardous waste	Hazard code
K007	Wastewater treatment sludge from the production of iron blue pigments	(T)
K008	Oven residue from the production of chrome oxide green pigments	(T)
Organic chemicals:		
K009	Distillation bottoms from the production of acetaldehyde from ethylene	(T)
K010	Distillation side cuts from the production of acetaldehyde from ethylene	(T)
K011	Bottom stream from the wastewater stripper in the production of acrylonitrile	(R, T)
K013	Bottom stream from the acetonitrile column in the production of acrylonitrile	(R, T)
K014	Bottoms from the acetonitrile purification column in the production of acrylonitrile	(T)
K015	Still bottoms from the distillation of benzyl chloride	(T)
K016	Heavy ends or distillation residues from the production of carbon tetrachloride	(T)
K017	Heavy ends (still bottoms) from the purification column in the production of epichlorohydrin.	(T)
K018	Heavy ends from the fractionation column in ethyl chloride production	(T)
K019	Heavy ends from the distillation of ethylene dichloride in ethylene dichloride production.	(T)
K020	Heavy ends from the distillation of vinyl chloride in vinyl chloride monomer production.	(T)
K021	Aqueous spent antimony catalyst waste from fluoromethanes production	(T)
K022	Distillation bottom tars from the production of phenol/acetone from cumene	(T)
K023	Distillation light ends from the production of phthalic anhydride from naphthalene	(T)
K024	Distillation bottoms from the production of phthalic anhydride from naphthalene	(T)
K093	Distillation light ends from the production of phthalic anhydride from ortho-xylene	(T)
K094	Distillation bottoms from the production of phthalic anhydride from ortho-xylene	(T)
K025	Distillation bottoms from the production of nitrobenzene by the nitration of benzene	(T)
K026	Stripping still tails from the production of methy ethyl pyridines	(T)
K027	Centrifuge and distillation residues from toluene diisocyanate production	(R, T)
K028	Spent catalyst from the hydrochlorinator reactor in the production of 1,1,1-trichloroethane.	(T)
K029	Waste from the product steam stripper in the production of 1,1,1-trichloroethane	(T)
K095	Distillation bottoms from the production of 1,1,1-trichloroethane	(T)
K096	Heavy ends from the heavy ends column from the production of 1,1,1-trichloroethane.	(T)
K030	Column bottoms or heavy ends from the combined production of trichloroethylene and perchloroethylene.	(T)
K083	Distillation bottoms from aniline production	(T)
K103	Process residues from aniline extraction from the production of aniline	(T)
K104	Combined wastewater streams generated from nitrobenzene/aniline production	(T)
K085	Distillation or fractionation column bottoms from the production of chlorobenzenes	(T)
K105	Separated aqueous stream from the reactor product washing step in the production of chlorobenzenes.	(T)
K111	Product washwaters from the production of dinitrotoluene via nitration of toluene	(C,T)
K112	Reaction by-product water from the drying column in the production of toluenediamine via hydrogenation of dinitrotoluene.	(T)
K113	Condensed liquid light ends from the purification of toluenediamine in the production of toluenediamine via hydrogenation of dinitrotoluene.	(T)
K114	Vicinals from the purification of toluenediamine in the production of toluenediamine via hydrogenation of dinitrotoluene.	(T)
K115	Heavy ends from the purification of toluenediamine in the production of toluenediamine via hydrogenation of dinitrotoluene.	(T)
K116	Organic condensate from the solvent recovery column in the production of toluene diisocyanate via phosgenation of toluenediamine.	(T)
K117	Wastewater from the reactor vent gas scrubber in the production of ethylene dibromide via bromination of ethene.	(T)
K118	Spent adsorbent solids from purification of ethylene dibromide in the production of ethylene dibromide via bromination of ethene.	(T)
K136	Still bottoms from the purification of ethylene dibromide in the production of ethylene dibromide via bromination of ethene.	(T)
Inorganic chemicals:		
K071	Brine purification muds from the mercury cell process in chlorine production, where separately prepurified brine is not used.	(T)
K073	Chlorinated hydrocarbon waste from the purification step of the diaphragm cell process using graphite anodes in chlorine production.	(T)
K106	Wastewater treatment sludge from the mercury cell process in chlorine production	(T)
Pesticides:		
K031	By-product salts generated in the production of MSMA and cacodylic acid	(T)
K032	Wastewater treatment sludge from the production of chlordane	(T)
K033	Wastewater and scrub water from the chlorination of cyclopentadiene in the production of chlordane.	(T)
K034	Filter solids from the filtration of hexachlorocyclopentadiene in the production of chlordane.	(T)
K097	Vacuum stripper discharge from the chlordane chlorinator in the production of chlordane.	(T)
K035	Wastewater treatment sludges generated in the production of creosote	(T)
K036	Still bottoms from toluene reclamation distillation in the production of disulfoton	(T)
K037	Wastewater treatment sludges from the production of disulfoton	(T)
K038	Wastewater from the washing and stripping of phorate production	(T)
K039	Filter cake from the filtration of diethylphosphorodithioic acid in the production of phorate.	(T)

Environmental Protection Agency § 261.33

Industry and EPA hazardous waste No.	Hazardous waste	Hazard code
K040	Wastewater treatment sludge from the production of phorate	(T)
K041	Wastewater treatment sludge from the production of toxaphene	(T)
K098	Untreated process wastewater from the production of toxaphene	(T)
K042	Heavy ends or distillation residues from the distillation of tetrachlorobenzene in the production of 2,4,5-T.	(T)
K043	2,6-Dichlorophenol waste from the production of 2,4-D	(T)
K099	Untreated wastewater from the production of 2,4-D	(T)
Explosives:		
K044	Wastewater treatment sludges from the manufacturing and processing of explosives	(R)
K045	Spent carbon from the treatment of wastewater containing explosives	(R)
K046	Wastewater treatment sludges from the manufacturing, formulation and loading of lead-based initiating compounds.	(T)
K047	Pink/red water from TNT operations	(R)
Petroleum refining:		
K048	Dissolved air flotation (DAF) float from the petroleum refining industry	(T)
K049	Slop oil emulsion solids from the petroleum refining industry	(T)
K050	Heat exchanger bundle cleaning sludge from the petroleum refining industry	(T)
K051	API separator sludge from the petroleum refining industry	(T)
K052	Tank bottoms (leaded) from the petroleum refining industry	(T)
Iron and steel:		
K061	Emission control dust/sludge from the primary production of steel in electric furnaces.	(T)
K062	Spent pickle liquor generated by steel finishing operations of plants that produce iron or steel.	(C,T)
Secondary lead:		
K069	Emission control dust/sludge from secondary lead smelting	(T)
K100	Waste leaching solution from acid leaching of emission control dust/sludge from secondary lead smelting.	(T)
Veterinary pharmaceuticals:		
K084	Wastewater treatment sludges generated during the production of veterinary pharmaceuticals from arsenic or organo-arsenic compounds.	(T)
K101	Distillation tar residues from the distillation of aniline-based compounds in the production of veterinary pharmaceuticals from arsenic or organo-arsenic compounds.	(T)
K102	Residue from the use of activated carbon for decolorization in the production of veterinary pharmaceuticals from arsenic or organo-arsenic compounds.	(T)
Ink formulation: K086	Solvent washes and sludges, caustic washes and sludges, or water washes and sludges from cleaning tubs and equipment used in the formulation of ink from pigments, driers, soaps, and stabilizers containing chromium and lead.	(T)
Coking:		
K060	Ammonia still lime sludge from coking operations	(T)
K087	Decanter tank tar sludge from coking operations	(T)

§§261.33 Discarded commercial chemical products, off-specification species, container residues, and spill residues thereof.

The following materials or items are hazardous wastes if and when they are discarded or intended to be discarded, when they are mixed with waste oil or used oil or other material and applied to the land for dust suppression or road treatment, or when, in lieu of their original intended use, they are produced for use as (or as a component of) a fuel, distributed for use as a fuel, or burned as a fuel.

(a) Any commercial chemical product, or manufacturing chemical intermediate having the generic name listed in paragraph (e) or (f) of this section.

(b) Any off-specification commercial chemical product or manufacturing chemical intermediate which, if it met specifications, would have the generic name listed in paragraph (e) or (f) of this section.

(c) Any container or inner liner removed from a container that has been used to hold any commercial chemical product or manufacturing chemical intermediate having the generic names listed in paragraph (e) of this section, or any container or inner liner removed from a container that has been used to hold any off-specification chemical produce and manufacturing chemical intermediate which, if it met specifications would have the generic name listed in paragraph (e) of this section, unless the container is empty as defined in § 261.7(b)(3) of this chapter.
[*Comment*: Unless the residue is being beneficially used or reused, or legitimately recycled or reclaimed; or being accumulated, stored, transported or treated prior to such use, re-use recycling or reclamation, EPA considers the residue to be intended for discard, and thus a hazardous waste. An example of a legitimate re-use of the residue would be where the residue remains in the container and the container is used to hold the same commercial chemical product or manufacturing chemical product or manufacturing chemical product or manufacturing chemical intermediate if previously held. An example of the discard of the residue would be where the drum is sent to a drum reconditioner who reconditions the drum but discards the residue.]

(d) Any residue or contaminated soil, water or other debris resulting from the cleanup of a spill into or on any land or water of any commerical chemical product or manufacturing chemical intermediate having the generic name listed in paragraph (e) or (f) of this section, or any residue or contaminated soil, water or other debris resulting from the cleanup of a spill, into or on any land or water, of any off-specification chemical intermediate which, if it met specifications, would have the generic name listed in paragraph (e) or (f) of this section.
[*Comment*: The phrase "commercial chemical product or manufacturing chemical intermediate having the generic name listed in . . ." refers to a chemical substance which is manufactured or formulated for commercial or manufacturing use which consists of the commercially pure grade of the chemical any technical grades of the chemical that are produced or marketed, and all formulations in which the chemical is the sole active ingredient. It does not refer to a material, such as a manufacturing process waste, that contains any of the substances listed in paragraph (e) or (f). Where a manufacturing process waste is deemed to be a hazardous waste because it contains a substance listed in paragraph (e) or (f), such waste will be listed in either § 261.31 or § 261.32 or will be identified as a hazardous waste by the characteristics set forth in Subpart C of this part.]

(e) The commercial chemical products, manufacturing chemical intermediates or off-specification commercial chemical products or manufacturing chemical intermediates referred to in paragraphs (a) through (d) of this section, are identified as acute hazardous wastes (H) and are subject to be the small quantity exclusion defined in § 261.5(e).

[*Comment*: For the convenience of the regulated community the primary hazardous properties of these materials have been indicated by the letters T (Toxicity), and R (Reactivity). Absence of a letter indicates that the compound only is listed for acute toxicity.]

These wastes and their corresponding EPA Hazardous Waste Numbers are:

Hazardous waste No.	Substance
P023	Acetaldehyde, chloro-
P002	Acetamide, N-(aminothioxomethyl)-
P057	Acetamide, 2-fluoro-
P058	Acetic acid, fluoro-, sodium salt
P066	Acetimidic acid, N-[(methylcarbamoyl)oxy]thio-, methyl ester
P001	3-(alpha-Acetonylbenzyl)-4-hydroxycoumarin and salts, when present at concentrations greater than 0.3%
P002	1-Acetyl-2-thiourea
P003	Acrolein
P070	Aldicarb
P004	Aldrin
P005	Allyl alcohol
P006	Aluminum phosphide
P007	5-(Aminomethyl)-3-isoxazolol
P008	4-aAminopyridine
P009	Ammonium picrate (R)
P119	Ammonium vanadate
P010	Arsenic acid
P012	Arsenic (III) oxide
P011	Arsenic (V) oxide
P011	Arsenic pentoxide
P012	Arsenic trioxide
P038	Arsine, diethyl-
P054	Aziridine
P013	Barium cyanide
P024	Benzenamine, 4-chloro-
P077	Benzenamine, 4-nitro-
P028	Benzene, (chloromethyl)-
P042	1,2-Benzenediol, 4-[1-hydroxy-2-(methylamino)ethyl]-
P014	Benzenethiol
P028	Benzyl chloride
P015	Beryllium dust
P016	Bis(chloromethyl) ether
P017	Bromoacetone
P018	Brucine
P021	Calcium cyanide
P123	Camphene, octachloro-
P103	Carbamimidoselenoic acid
P022	Carbon bisulfide
P022	Carbon disulfide
P095	Carbonyl chloride
P033	Chlorine cyanide
P023	Chloroacetaldehyde
P024	p-Chloroaniline
P026	1-(o-Chlorophenyl)thiourea
P027	3-Chloropropionitrile
P029	Copper cyanides
P030	Cyanides (soluble cyanide salts), not elsewhere specified
P031	Cyanogen
P033	Cyanogen chloride
P036	Dichlorophenylarsine
P037	Dieldrin
P038	Diethylarsine
P039	O,O-Diethyl S-[2-(ethylthio)ethyl] phosphorodithioate
P041	Diethyl-p-nitrophenyl phosphate
P040	O,O-Diethyl O-pyrazinyl phosphorothioate
P043	Diisopropyl fluorophosphate
P044	Dimethoate
P045	3,3-Dimethyl-1-(methylthio)-2-butanone, O-[(methylamino)carbonyl] oxime
P071	O,O-Dimethyl O-p-nitrophenyl phosphorothioate
P082	Dimethylnitrosamine
P046	alpha, alpha-Dimethylphenethylamine
P047	4,6-Dinitro-o-cresol and salts
P034	4,6-Dinitro-o-cyclohexylphenol
P048	2,4-Dinitrophenol
P020	Dinoseb
P085	Diphosphoramide, octamethyl-
P039	Disulfoton
P049	2,4-Dithiobiuret
P109	Dithiopyrophosphoric acid, tetraethyl ester
P050	Endosulfan
P088	Endothall
P051	Endrin
P042	Epinephrine
P046	Ethanamine, 1,1-dimethyl-2-phenyl-
P084	Ethenamine, N-methyl-N-nitroso-
P101	Ethyl cyanide
P054	Ethylenimine
P097	Famphur
P056	Fluorine
P057	Fluoroacetamide
P058	Fluoroacetic acid, sodium salt
P065	Fulminic acid, mercury(II) salt (R,T)
P059	Heptachlor
P051	1,2,3,4,10,10-Hexachloro-6,7-epoxy-1,4,4a,5,6,7,8,8a-octahydro-endo,endo-1,4:5,8-dimethanonaphthalene
P037	1,2,3,4,10,10-Hexachloro-6,7-epoxy-1,4,4a,5,6,7,8,8a-octahydro-endo,exo-1,4:5,8-demethanonaphthalene
P060	1,2,3,4,10,10-Hexachloro-1,4,4a,5,8,8a-hexahydro-1,4:5,8-endo, endo-dimeth- anonaphthalene
P004	1,2,3,4,10,10-Hexachloro-1,4,4a,5,8,8a-hexahydro-1,4:5,8-endo,exo-dimethanonaphthalene
P060	Hexachlorohexahydro-exo,exo-dimethanonaphthalene
P062	Hexaethyl tetraphosphate
P116	Hydrazinecarbothioamide
P068	Hydrazine, methyl-
P063	Hydrocyanic acid
P063	Hydrogen cyanide

Hazardous waste No.	Substance
P096	Hydrogen phosphide
P064	Isocyanic acid, methyl ester
P007	3(2H)-Isoxazolone, 5-(aminomethyl)-
P092	Mercury, (acetato-O)phenyl-
P065	Mercury fulminate (R,T)
P016	Methane, oxybis(chloro-
P112	Methane, tetranitro- (R)
P118	Methanethiol, trichloro-
P059	4,7-Methano-1H-indene, 1,4,5,6,7,8,8-heptachloro-3a,4,7,7a-tetrahydro-
P066	Methomyl
P067	2-Methylaziridine
P068	Methyl hydrazine
P064	Methyl isocyanate
P069	2-Methyllactonitrile
P071	Methyl parathion
P072	alpha-Naphthylthiourea
P073	Nickel carbonyl
P074	Nickel cyanide
P074	Nickel(II) cyanide
P073	Nickel tetracarbonyl
P075	Nicotine and salts
P076	Nitric oxide
P077	p-Nitroaniline
P078	Nitrogen dioxide
P076	Nitrogen(II) oxide
P078	Nitrogen(IV) oxide
P081	Nitroglycerine (R)
P082	N-Nitrosodimethylamine
P084	N-Nitrosomethylvinylamine
P050	5-Norbornene-2,3-dimethanol, 1,4,5,6,7,7-hexachloro, cyclic sulfite
P085	Octamethylpyrophosphoramide
P087	Osmium oxide
P087	Osmium tetroxide
P088	7-Oxabicyclo[2.2.1]heptane-2,3-dicarboxylic acid
P089	Parathion
P034	Phenol, 2-cyclohexyl-4,6-dinitro-
P048	Phenol, 2,4-dinitro-
P047	Phenol, 2,4-dinitro-6-methyl-
P020	Phenol, 2,4-dinitro-6-(1-methylpropyl)-
P009	Phenol, 2,4,6-trinitro-, ammonium salt (R)
P036	Phenyl dichloroarsine
P092	Phenylmercuric acetate
P093	N-Phenylthiourea
P094	Phorate
P095	Phosgene
P096	Phosphine
P041	Phosphoric acid, diethyl p-nitrophenyl ester
P044	Phosphorodithioic acid, O,O-dimethyl S-[2-(methylamino)-2-oxoethyl]ester
P043	Phosphorofluoric acid, bis(1-methylethyl)-ester
P094	Phosphorothioic acid, O,O-diethyl S-(ethylthio)methyl ester
P089	Phosphorothioci acid, O,O-diethyl O-(p-nitrophenyl) ester
P040	Phosphorothioic acid, O,O-diethyl O- pyrazinyl ester
P097	Phosphorothioic acid, O,O-dimethyl O-[p-((dimethylamino)-sulfonyl)phenyl]ester

Hazardous waste No.	Substance
P110	Plumbane, tetraethyl-
P098	Potassium cyanide
P099	Potassium silver cyanide
P070	Propanal, 2-methyl-2-(methylthio)-, O-[(methylamino)carbonyl]oxime
P101	Propanenitrile
P027	Propanenitrile, 3-chloro-
P069	Propanenitrile, 2-hydroxy-2-methyl-
P081	1,2,3-Propanetriol, trinitrate- (R)
P017	2-Propanone, 1-bromo-
P102	Propargyl alcohol
P003	2-Propenal
P005	2-Propen-1-ol
P067	1,2-Propylenimine
P102	2-Propyn-1-ol
P008	4-Pyridinamine
P075	Pyridine, (S)-3-(1-methyl-2-pyrrolidinyl)-, and salts
P111	Pyrophosphoric acid, tetraethyl ester
P103	Selenourea
P104	Silver cyanide
P105	Sodium azide
P106	Sodium cyanide
P107	Strontium sulfide
P108	Strychnidin-10-one, and salts
P018	Strychnidin-10-one, 2,3-dimethoxy-
P108	Strychnine and salts
P115	Sulfuric acid, thallium(I) salt
P109	Tetraethyldithiopyrophosphate
P110	Tetraethyl lead
P111	Tetraethylpyrophosphate
P112	Tetranitromethane (R)
P062	Tetraphosphoric acid, hexaethyl ester
P113	Thallic oxide
P113	Thallium(III) oxide
P114	Thallium(I) selenite
P115	Thallium(I) sulfate
P045	Thiofanox
P049	Thioimidodicarbonic diamide
P014	Thiophenol
P116	Thiosemicarbazide
P026	Thiourea, (2-chlorophenyl)-
P072	Thiourea, 1-naphthalenyl-
P093	Thiourea, phenyl-
P123	Toxaphene
P118	Trichloromethanethiol
P119	Vanadic acid, ammonium salt
P120	Vanadium pentoxide
P120	Vanadium(V) oxide
P001	Warfarin, when present at concentrations greater than 0.3%
P121	Zinc cyanide
P122	Zinc phosphide (R,T)
P122	Zinc phosphide, when present at concentrations greater than 10%

(f) The commercial chemical products, manufacturing chemical intermediates, or off-specification commercial chemical products referred to in paragraphs (a) through (d) of this section, are identified as toxic wastes (T), unless otherwise designated and are subject to the small quantity generator exclusion defined in § 261.5 (a) and (g). [*Comment*: For the convenience of the regulated community, the primary hazardous properties of these materials have been indicated by the letters T (Toxicity), R (Reactivity), I (Ignitability) and C (Corrosivity). Absence of a letter indicates that the compound is only listed for toxicity.]

These wastes and their corresponding EPA Hazardous Waste Numbers are:

Hazardous Waste No.	Substance
U001	Acetaldehyde (I)
U034	Acetaldehyde, trichloro-
U187	Acetamide, N-(4-ethoxyphenyl)-
U005	Acetamide, N-9H-fluoren-2-yl-
U112	Acetic acid, ethyl ester (I)
U144	Acetic acid, lead salt
U214	Acetic acid, thallium(I) salt
U002	Acetone (I)
U003	Acetonitrile (I,T)
U248	3-(alpha-Acetonylbenzyl)-4-hydroxycoumarin and salts, when present at concentrations of 0.3% or less
U004	Acetophenone
U005	2-Acetylaminofluorene
U006	Acetyl chloride (C,R,T)
U007	Acrylamide
U008	Acrylic acid (I)
U009	Acrylonitrile
U150	Alanine, 3-[p-bis(2-chloroethyl)amino] phenyl-, L-
U328	2-Amino-l-methylbenzene
U353	4-Amino-l-methylbenzene
U011	Amitrole
U012	Aniline (I,T)
U014	Auramine
U015	Azaserine
U010	Azirino(2',3':3,4)pyrrolo(1,2-a)indole-4,7-dione, 6-amino-8-[((aminocarbonyl) oxy)methyl]-1,1a,2,8,8a,8b-hexahydro-8a-methoxy-5-methyl-,
U157	Benz[j]aceanthrylene, 1,2-dihydro-3-methyl-
U016	Benz[c]acridine
U016	3,4-Benzacridine
U017	Benzal chloride
U018	Benz[a]anthracene
U018	1,2-Benzanthracene
U094	1,2-Benzanthracene, 7,12-dimethyl-
U012	Benzenamine (I,T)
U014	Benzenamine, 4,4'-carbonimidoylbis(N,N-dimethyl-
U049	Benzenamine, 4-chloro-2-methyl-
U093	Benzenamine, N,N'-dimethyl-4-phenylazo-
U158	Benzenamine, 4,4'-methylenebis(2-chloro-
U222	Benzenamine, 2-methyl-, hydrochloride
U181	Benzenamine, 2-methyl-5-nitro
U019	Benzene (I,T)
U038	Benzeneacetic acid, 4-chloro-alpha-(4-chlorophenyl)-alpha-hydroxy, ethyl ester
U030	Benzene, 1-bromo-4-phenoxy-
U037	Benzene, chloro-
U190	1,2-Benzenedicarboxylic acid anhydride
U028	1,2-Benzenedicarboxylic acid, [bis(2-ethylhexyl)] ester

Hazardous Waste No.	Substance
U069	1,2-Benzenedicarboxylic acid, dibutyl ester
U088	1,2-Benzenedicarboxylic acid, diethyl ester
U102	1,2-Benzenedicarboxylic acid, dimethyl ester
U107	1,2-Benzenedicarboxylic acid, di-n-octyl ester
U070	Benzene, 1,2-dichloro-
U071	Benzene, 1,3-dichloro-
U072	Benzene, 1,4-dichloro-
U017	Benzene, (dichloromethyl)-
U223	Benzene, 1,3-diisocyanatomethyl- (R,T)
U239	Benzene, dimethyl-(I,T)
U201	1,3-Benzenediol
U127	Benzene, hexachloro-
U056	Benzene, hexahydro- (I)
U188	Benzene, hydroxy-
U220	Benzene, methyl-
U105	Benzene, 1-methyl-1-2,4-dinitro-
U106	Benzene, 1-methyl-2,6-dinitro-
U203	Benzene, 1,2-methylenedioxy-4-allyl-
U141	Benzene, 1,2-methylenedioxy-4-propenyl-
U090	Benzene, 1,2-methylenedioxy-4-propyl-
U055	Benzene, (1-methylethyl)- (I)
U169	Benzene, nitro- (I,T)
U183	Benzene, pentachloro-
U185	Benzene, pentachloro-nitro-
U020	Benzenesulfonic acid chloride (C,R)
U020	Benzenesulfonyl chloride (C,R)
U207	Benzene, 1,2,4,5-tetrachloro-
U023	Benzene, (trichloromethyl)-(C,R,T)
0234	Benzene, 1,3,5-trinitro- (R,T)
U021	Benzidine
U202	1,2-Benzisothiazolin-3-one, 1,1-dioxide
U120	Benzo[j,k]fluorene
U022	Benzo[a]pyrene
U022	3,4-Benzopyrene
U197	p-Benzoquinone
U023	Benzotrichloride (C,R,T)
U050	1,2-Benzphenanthrene
U085	2,2'-Bioxirane (I,T)
U021	(1,1'-Biphenyl)-4,4'-diamine
U073	(1,1'-Biphenyl)-4,4'-diamine, 3,3'-dichloro-
U091	(1,1'-Biphenyl)-4,4'-diamine, 3,3'-dimethoxy-
U095	(1,1'-Biphenyl)-4,4'-diamine, 3,3'-dimethyl-
U024	Bis(2-chloroethoxy) methane
U027	Bis(2-chloroisopropyl) ether
U244	Bis(dimethylthiocarbamoyl) disulfide
U028	Bis(2-ethylhexyl) phthalate
U246	Bromine cyanide
U225	Bromoform
U030	4-Bromophenyl phenyl ether
U128	1,3-Butadiene, 1,1,2,3,4,4-hexachloro-
U172	1-Butanamine, N-butyl-N-nitroso-
U035	Butanoic acid, 4-[Bis(2-chloroethyl)amino] benzene-
U031	1-Butanol (I)

Hazardous Waste No.	Substance
U159	2-Butanone (I,T)
U160	2-Butanone peroxide (R,T)
U053	2-Butenal
U074	2-Butene, 1,4-dichloro- (I,T)
U031	n-Butyl alchohol (I)
U136	Cacodylic acid
U032	Calcium chromate
U238	Carbamic acid, ethyl ester
U178	Carbamic acid, methylnitroso-, ethyl ester
U176	Carbamide, N-ethyl-N-nitroso-
U177	Carbamide, N-methyl-N-nitroso-
U219	Carbamide, thio-
U097	Carbamoyl chloride, dimethyl-
U215	Carbonic acid, dithallium(I) salt
U156	Carbonochloridic acid, methyl ester (I,T)
U033	Carbon oxyfluoride (R,T)
U211	Carbon tetrachloride
U033	Carbonyl fluoride (R,T)
U034	Chloral
U035	Chlorambucil
U036	Chlordane, technical
U026	Chlornaphazine
U037	Chlorobenzene
U039	4-Chloro-m-cresol
U041	1-Chloro-2,3-epoxypropane
U042	2-Chloroethyl vinyl ether
U044	Chloroform
U046	Chloromethyl methyl ether
U047	beta-Chloronaphthalene
U048	o-Chlorophenol
U049	4-Chloro-o-toluidine, hydrochloride
U032	Chromic acid, calcium salt
U050	Chrysene
U051	Creosote
U052	Cresols
U052	Cresylic acid
U053	Crotonaldehyde
U055	Cumene (I)
U246	Cyanogen bromide
U197	1,4-Cyclohexadienedione
U056	Cyclohexane (I)
U057	Cyclohexanone (I)
U130	1,3-Cyclopentadiene, 1,2,3,4,5,5-hexa- chloro-
U058	Cyclophosphamide
U240	2,44-D, salts and esters
U059	Daunomycin
U060	DDD
U061	DDT
U142	Decachlorooctahydro-1,3,4-metheno-2H-cyclobuta[c,d]-pentalen-2-one
U062	Diallate
U133	Diamine (R,T)
U221	Diaminotoluene
U063	Dibenz[a,h]anthracene
U063	1,2:5,6-Dibenzanthracene
U064	1,2:7,8-Dibenzopyrene
U064	Dibenz[a,i]pyrene
U066	1,2-Dibromo-3-chloropropane
U069	Dibutyl phthalate
U062	S-(2,3-Dichloroallyl) diisopropylthiocarbamate
U070	o-Dichlorobenzene
U071	m-Dichlorobenzene
U072	p-Dichlorobenzene
U073	3,3'-Dichlorobenzidine
U074	1,4-Dichloro-2-butene (I,T)
U075	Dichlorodifluoromethane
U192	3,5-Dichloro-N-(1,1-dimethyl-2-propynyl) benzamide
U060	Dichloro diphenyl dichloroethane
U061	Dichloro diphenyl trichloroethane
U078	1,1-Dichloroethylene
U079	1,2-Dichloroethylene
U025	Dichloroethyl ether
U081	2,4-Dichlorophenol
U082	2,6-Dichlorophenol
U240	2,4-Dichlorophenoxyacetic acid, salts and esters
U083	1,2-Dichloropropane
U084	1,3-Dichloropropene
U085	1,2:3,4-Diepoxybutane (I,T)
U108	1,4-Diethylene dioxide

Hazardous Waste No.	Substance
U086	N,N-Diethylhydrazine
U087	O,O-Diethyl-S-methyl-dithiophosphate
U088	Diethyl phthalate
U089	Diethylstilbestrol
U148	1,2-Dihydro-3,6-pyradizinedione
U090	Dihydrosafrole
U091	3,3'-Dimethoxybenzidine
U092	Dimethylamine (I)
U093	Dimethylaminoazobenzene
U094	7,12-Dimethylbenz[a]anthracene
U095	3,3'-Dimethylbenzidine
U096	alpha,alpha-Dimethylbenzylhydroperoxide (R)
U097	Dimethylcarbamoyl chloride
U098	1,1-Dimethylhydrazine
U099	1,2-Dimethylhydrazine
U101	2,4-Dimethylphenol
U102	Dimethyl phthalate
U103	Dimethyl sulfate
U105	2,4-Dinitrotoluene
U106	2,6-Dinitrotoluene
U107	Di-n-octyl phthalate
U108	1,4-Dioxane
U109	1,2- Diphenylhydrazine
U110	Dipropylamine (I)
U111	Di-N-propylnitrosamine
U001	Ethanal (I)
U174	Ethanamine, N-ethyl-N-nitroso-
U067	Ethane, 1,2-dibromo-
U076	Ethane, 1,1-dichloro-
U077	Ethane, 1,2-dichloro-
U114	1,2-Ethanediylbiscarbamodithioic acid
U131	Ethane, 1,1,1,2,2,2-hexachloro-
U024	Ethane, 1,1'-[methylenebis(oxy)]bis[2-chloro-
U003	Ethanenitrile (I, T)
U117	Ethane,1,1'-oxybis- (I)
U025	Ethane, 1,1'-oxybis[2-chloro-
U184	Ethane, pentachloro-
U208	Ethane, 1,1,1,2-tetrachloro-
U209	Ethane, 1,1,2,2-tetrachloro-
U218	Ethanethioamide
U247	Ethane, 1,1,1,-trichloro-2,2-bis(p-methoxy-phenyl).
U227	Ethane, 1,1,2-trichloro-
U043	Ethene, chloro-
U042	Ethene, 2-chloroethoxy-
U078	Ethene, 1,1-dichloro-
U079	Ethene, trans-1,2-dichloro-
U210	Ethene, 1,1,2,2-tetrachloro-
U173	Ethanol, 2,2'-(nitrosoimino)bis-
U004	Ethanone, 1-phenyl-
U006	Ethanoyl chloride (C,R,T)
U359	2-Ethoxyethanol.
U112	Ethyl acetate (I)
U113	Ethyl acrylate (I)
U238	Ethyl carbamate (urethan)
U038	Ethyl 4,4'-dichlorobenzilate
U359	Ethylene glycol monoethyl ether.
U114	Ethylenebis(dithiocarbamic acid)
U067	Etylene dibromide
U077	Ethylene dichloride
U115	Ethlene oxide (I,T)
U116	Ethylene thiourea
U117	Ethyl ether (I)
U076	Ethylidene dichloride
U118	Ethylmethacrylate
U119	Ethyl methanesulfonate
U139	Ferric dextran
U120	Fluoranthene
U122	Formaldehyde
U123	Formic acid (C,T)
U124	Furan (I)
U125	2-Furancarboxaldehyde (I)
U147	2,5-Furandione
U213	Furan, tetrahydro- (I)
U125	Furfural (I)
U124	Furfuran (I)
U206	D-Glucopyranose, 2-deoxy-2(3-methyl-3-nitro-soureido)-
U126	Glycidylaldehyde
U163	Guanidine, N-nitroso-N-methyl-N'nitro-

Hazardous Waste No.	Substance
U127	Hexachlorobenzene
U128	Hexachlorobutadiene
U129	Hexachlorocyclohexane (gamma isomer)
U130	Hexachlorocyclopentadiene
U131	Hexachloroethane
U132	Hexachlorophene
U243	Hexachloropropene
U133	Hydrazine (R,T)
U086	Hydrazine, 1,2-diethyl-
U098	Hydrazine, 1,1-dimethyl-
U099	Hydrazine, 1,2-dimethyl-
U109	Hydrazine, 1,2-diphenyl-
U134	Hydrofluoric acid (C,T)
U134	Hydrogen fluoride (C,T)
U135	Hydrogen sulfide
U096	Hydroperoxide, 1-methyl-1-phenylethyl- (R)
U136	Hydroxydimethylarsine oxide
U116	2-Imidazolidinethione
U137	Indeno[1,2,3-cd]pyrene
U139	Iron dextran
U140	Isobutyl alcohol (I,T)
U141	Isosafrole
U142	Kepone
U143	Lasiocarpine
U144	Lead acetate
U145	Lead phosphate
U146	Lead subacetate
U129	Lindane
U147	Maleic anhydride
U148	Maleic hydrazide
U149	Malononitrile
U150	Melphalan
U151	Mercury
U152	Methacrylonitrile (I,T)
U092	Methanamine, N-methyl- (I)
U029	Methane, bromo-
U045	Methane, chloro- (I,T)
U046	Methane, chloromethoxy-
U068	Methane, dibromo-
U080	Methane, dichloro-
U075	Methane, dichlorodifluoro-
U138	Methane, iodo-
U119	Methanesulfonic acid, ethyl ester
U211	Methane, tetrachloro-
U121	Methane, trichlorofluoro-
U153	Methanethiol (I,T)
U225	Methane, tribromo-
U044	Methane, trichloro-
U121	Methane, trichlorofluoro-
U123	Methanoic acid (C,T)
U036	4,7-Methanoindan, 1,2,4,5,6,7,8,8-octachloro-3a,4,7,7a-tetrahydro-
U154	Methanol (I)
U155	Methapyrilene
U247	Methoxychlor.
U154	Methyl alcohol (I)
U029	Methyl bromide
U186	1-Methylbutadiene (I)
U045	Methyl chloride (I,T)
U156	Methyl chlorocarbonate (I,T)
U226	Methylchloroform
U157	3-Methylcholanthrene
U158	4,4'-Methylenebis(2-chloroaniline)
U132	2,2'-Methylenebis(3,4,6-trichlorophenol)
U068	Methylene bromide
U080	Methylene chloride
U122	Methylene oxide
U159	Methyl ethyl ketone (I,T)
U160	Methyl ethyl ketone peroxide (R,T)
U138	Methyl iodide
U161	Methyl isobutyl ketone (I)
U162	Methyl methacrylate (I,T)
U163	N-Methyl-N'-nitro-N-nitrosoguanidine
U161	4-Methyl-2-pentanone (I)
U164	Methylthiouracil
U010	Mitomycin C
U059	5,12-Naphthacenedione, (8S-cis)-8-acetyl-10-[(3-amino-2,3,6-trideoxy-alpha-L-lyxo-hexopyranosyl)oxyl]-7,8,9,10-tetrahydro-6,8,11-trihydroxy-1-methoxy-
U165	Naphthalene
U047	Naphthalene, 2-chloro-
U166	1,4-Naphthalenedione
U236	2,7-Naphthalenedisulfonic acid, 3,3'-[(3,3'-dimethyl-(1,1'-biphenyl)-4,4'diyl)]-bis (azo)bis(5-amino-4-hydroxy)-,tetrasodium salt
U166	1,4,Naphthaquinone
U167	1-Naphthylamine
U168	2-Naphthylamine
U167	alpha-Naphthylamine
U168	beta-Naphthylamine
U026	2-Naphthylamine, N,N'-bis(2-chloromethyl)-
U169	Nitrobenzene (I,T)
U170	p-Nitrophenol
U171	2-Nitropropane (I,T)
U172	N-Nitrosodi-n-butylamine
U173	N-Nitrosodiethanolamine
U174	N-Nitrosodiethylamine
U111	N-Nitroso-N-propylamine
U176	N-Nitroso-N-ethylurea
U177	N-Nitroso-N-methylurea
U178	N-Nitroso-N-methylurethane
U179	N-Nitrosopiperidine
U180	N-Nitrosopyrrolidine
U181	5-Nitro-o-toluidine
U193	1,2-Oxathiolane, 2,2-dioxide
U058	2H-1,3,2-Oxazaphosphorine, 2-[bis(2-chloroethyl)amino]tetrahydro-, oxide 2-
U115	Oxirane (I,T)
U041	Oxirane, 2-(chloromethyl)-
U182	Paraldehyde
U183	Pentachlorobenzene
U184	Pentachloroethane
U185	Pentachloronitrobenzene
See F027	Pentachlorophenol
U186	1,3-Pentadiene (I)
U187	Phenacetin
U188	Phenol
U048	Phenol, 2-chloro-
U039	Phenol, 4-chloro-3-methyl-
U081	Phenol, 2,4-dichloro-
U082	Phenol, 2,6-dichloro-
U101	Phenol, 2,4-dimethyl-
U170	Phenol, 4-nitro-
See F027	Phenol, pentachloro-
Do	Phenol, 2,3,4,6-tetrachloro-
Do	Phenol, 2,4,5-trichloro-
Do	Phenol, 2,4,6-trichloro-
U137	1,10-(1,2-phenylene)pyrene
U145	Phosphoric acid, Lead salt
U087	Phosphorodithioic acid, O,O-diethyl-, S-methylester
U189	Phosphorous sulfide (R)
U190	Phthalic anhydride
U191	2-Picoline
U192	Pronamide
U194	1-Propanamine (I,T)
U110	1-Propanamine, N-propyl- (I)
U066	Propane, 1,2-dibromo-3-chloro-
U149	Propanedinitrile
U171	Propane, 2-nitro- (I,T)
U027	Propane, 2,2'oxybis[2-chloro-
U193	1,3-Propane sultone
U235	1-Propanol, 2,3-dibromo-, phosphate (3:1)
U126	1-Propanol, 2,3-epoxy-
U140	1-Propanol, 2-methyl- (I,T)
U002	2-Propanone (I)
U007	2-Propenamide
U084	Propene, 1,3-dichloro-
U243	1-Propene, 1,1,2,3,3,3-hexachloro-
U009	2-Propenenitrile
U152	2-Propenenitrile, 2-methyl- (I,T)
U008	2-Propenoic acid (I)
U113	2-Propenoic acid, ethyl ester (I)
U118	2-Propenoic acid, 2-methyl-, ethyl ester
U162	2-Propenoic acid, 2-methyl-, methyl ester (I,T)
See F027	Propionic acid, 2-(2,4,5-trichlorophenoxy)-
U194	n-Propylamine (I,T)
U083	Propylene dichloride
U196	Pyridine
U155	Pyridine, 2-[(2-(dimethylamino)-2-thenylamino]-
U179	Pyridine, hexahydro-N-nitroso-

Hazardous Waste No.	Substance
U191	Pyridine, 2-methyl-
U164	4(1H)-Pyrimidinone, 2,3-dihydro-6-methyl-2-thioxo-
U180	Pyrrole, tetrahydro-N-nitroso-
U200	Reserpine
U201	Resorcinol
U202	Saccharin and salts
U203	Safrole
U204	Selenious acid
U204	Selenium dioxide
U205	Selenium disulfide (R,T)
U015	L-Serine, diazoacetate (ester)
See F027	Silvex
U089	4,4'-Stilbenediol, alpha,alpha'-diethyl-
U206	Streptozotocin
U135	Sulfur hydride
U103	Sulfuric acid, dimethyl ester
U189	Sulfur phosphide (R)
U205	Sulfur selenide (R,T)
See F027	2,4,5-T
U207	1,2,4,5-Tetrachlorobenzene
U208	1,1,1,2-Tetrachloroethane
U209	1,1,2,2-Tetrachloroethane
U210	Tetrachloroethylene
See F027	2,3,4,6-Tetrachlorophenol
U213	Tetrahydrofuran (I)
U214	Thallium(I) acetate
U215	Thallium(I) carbonate
U216	Thallium(I) chloride
U217	Thallium(I) nitrate
U218	Thioacetamide
U153	Thiomethanol (I,T)
U219	Thiourea
U244	Thiram
U220	Toluene
U221	Toluenediamine
U223	Toluene diisocyanate (R,T)
U328	*o*-Toluidine
U222	O-Toluidine hydrochloride
U353	*p*-Toluidine
U011	1H-1,2,4-Triazol-3-amine
U226	1,1,1-Trichloroethane
U227	1,1,2-Trichloroethane
U228	Trichloroethene
U228	Trichloroethylene
U121	Trichloromonofluoromethane
See F027	2,4,5-Trichlorophenol
Do	2,4,6-Trichlorophenol
Do	2,4,5-Trichlorophenoxyacetic acid
U234	sym-Trinitrobenzene (R,T)
U182	1,3,5-Trioxane, 2,4,5-trimethyl-
U235	Tris(2,3-dibromopropyl) phosphate
U236	Trypan blue
U237	Uracil, 5[bis(2-chloromethyl)amino]-
U237	Uracil mustard
U043	Vinyl chloride
U248	Warfarin, when present at concentrations of 0.3% or less
U239	Xylene (I)
U200	Yohimban-16-carboxylic acid, 11,17-dimethoxy-18-[(3,4,5-trimethoxy-benzoyl)oxy]-, methyl ester
U249	Zinc phosphide, when present at concentrations of 10% or less.

CHAPTER 1, APPENDIX III

Hazardous Waste Agencies and Small Quantity Limits

EPA REGIONAL OFFICES

EPA Region 1
State Waste Programs Branch
JFK Federal Building
Boston, MA 02203
(617) 223-3468
Connecticut, Massachusetts, Maine, New Hampshire, Rhode Island, Vermont

EPA Region 2
Air and Waste Management Division
26 Federal Plaza
New York, NY 10278
(212) 264–5175
New Jersey, New York, Puerto Rico, Virgin Islands

EPA Region 3
Waste Management Branch
841 Chestnut Street
Philadelphia, PA 19107
(215) 597-9336
Delaware, Maryland, Pennsylvania, Virgina, West Virginia, District of Columbia

EPA Region 4
Hazardous Waste Management Division
345 Courtland Street, N.E.
Atlanta, GA 30365
(404) 347-3016
Alabama, Florida, Georgia, Kentucky, Mississippi, North Carolina, South Carolina, Tennessee

EPA Region 5
RCRA Activities
230 S. Dearborn Street
Chicago, IL 60604
(312) 353-2000
Illinois, Indiana, Michigan, Minnesota, Ohio, Wisconsin

EPA Region 6
Air and Hazardous Materials Division
1201 Elm Street
Dallas, TX 75270
(214) 767-2600
Arkansas, Louisiana, New Mexico, Oklahoma, Texas

EPA Region 7
RCRA Branch
726 Minnesota Avenue
Kansas City, KS 66101
(913) 236-2800
Iowa, Kansas, Missouri, Nebraska

EPA Region 8
Waste Management Division (8HWM-ON)
One Denver Place
999 18th Street, Suite 1300
Denver, CO 80202–2413
(303) 293-1502
Colorado, Montana, North Dakota, South Dakota, Utah, Wyoming

EPA Region 9
Toxics and Waste Management Division
215 Fremont Street
San Francisco, CA 94105
(415) 974-7472
Arizona, California, Hawaii, Nevada, American Samoa, Guam, Trust Territories of the Pacific

EPA Region 10
Waste Management Branch—MS-530
1200 Sixth Avenue
Seattle, WA 98101
(206) 442-2777
Alaska, Idaho, Oregon, Washington

STATE AGENCIES*

Alabama
Alabama Department of Environmental Management
Land Division
1751 Federal Drive
Montgomery, AL 36130
(205) 271-7730
SMALL QUANTITY LIMITS: SAME AS FEDERAL

Alaska
Department of Environmental Conservation
P.O. Box O
Juneau, AK 99811
(907) 465-2666
SMALL QUANTITY LIMITS: SAME AS FEDERAL

*States which currently have fully authorized programs are marked with an asterisk.

Arizona*
Arizona Department of Health Services
Office of Waste and Water Quality
2005 N. Central Avenue, Room 304
Phoenix, AZ 85004
(602) 255-2211
SMALL QUANTITY LIMITS: SAME AS FEDERAL

Arkansas*
Department of Pollution Control and Ecology
Hazardous Waste Division
P.O. Box 9583
8001 National Drive
Little Rock, AR 72219
(501) 562-7444
SMALL QUANTITY LIMITS: SAME AS FEDERAL

California*
Department of Health Services
Toxic Substances Control Division
714 P Street, Room 1253
Sacramento, CA 95814
(916) 324-1826
SMALL QUANTITY LIMITS: NO EXEMPTIONS FOR SMALL QUANTITY GENERATORS

Colorado*
Colorado Department of Health
Waste Management Division
4210 E. 11th Avenue
Denver, CO 80220
(303) 320-8333 ext. 4364
SMALL QUANTITY LIMITS: SAME AS FEDERAL

Connecticut
Department of Environmental Protection
Hazardous Waste Management Section
State Office Building
165 Capitol Avenue
Hartford, CT 06106
(203) 566-8843
SMALL QUANTITY LIMITS: 1000 KG (NO ACUTE OR SPILL RESIDUE PROVISIONS)

Delaware*
Department of Natural Resources and Environmental Control
Waste Management Section
P.O. Box 1401
Dover, DE 19903
(302) 736-4781
SMALL QUANTITY LIMITS: SAME AS FEDERAL

District of Columbia
Department of Consumer and Regulatory Affairs
Pesticides and Hazardous Waste Materials Division
5010 Overlook Avenue, S.W., Room 114
Washington, DC 20032
(202) 767-8414
SMALL QUANTITY LIMITS: SAME AS FEDERAL

Florida*
Department of Environmental Regulation
Solid and Hazardous Waste Section
Twin Towers Office Building
2600 Blair Stone Road
Tallahassee, FL 32301
(904) 488-0300
SMALL QUANTITY LIMITS: SAME AS FEDERAL

Georgia*
Georgia Environmental Protection Division
Hazardous Waste Management Program
Land Protection Branch
Floyd Towers East, Suite 1154
205 Butler Street, S.E.
Atlanta, GA 30334
(404) 656-2833 or (800) 334-2373
SMALL QUANTITY LIMITS: SAME AS FEDERAL

Hawaii
Department of Health
Environmental Health Division
P.O. Box 3378
Honolulu, HI 96801
(808) 548-4383
SMALL QUANTITY LIMITS: SAME AS FEDERAL

Idaho
Department of Health and Welfare
Bureau of Hazardous Materials
450 W. State Street
Boise, ID 83720
(208) 334-5879
SMALL QUANTITY LIMITS: SAME AS FEDERAL

Illinois*
Environmental Protection Agency
Division of Land Pollution Control
2200 Churchill Road, #24
Springfield, IL 62706
(217) 782-6761
SMALL QUANTITY LIMITS: 100 KG FOR NONACUTE WASTE, 1 KG FOR HIGHLY ACUTE WASTE

Indiana*
Department of Environmental Management
Office of Solid and Hazardous Waste
105 S. Meridian
Indianapolis, IN 46225
(317) 232-4535
SMALL QUANTITY LIMITS: SAME AS FEDERAL

Iowa
U.S. EPA Region 7
Hazardous Materials Branch
726 Minnesota Avenue
Kansas City, KS 66101
(913) 236-2888 Iowa RCRA Toll Free: (800) 223-0425
SMALL QUANTITY LIMITS: SAME AS FEDERAL

Kansas*
Department of Health and Environment
Bureau of Waste Management
Forbes Field, Bldg. 321
Topeka, KS 66620
(913) 862-9360 ext. 292
SMALL QUANTITY LIMITS: 25 KG/ MONTH FOR NONACUTE WASTE AND SPILL RESIDUES, 1 KG/ MONTH FOR ACUTE WASTE

Kentucky*
Natural Resources and Environmental Protection Cabinet
Division of Waste Management
18 Reilly Road
Frankfort, KY 40601
(502) 564-6716
SMALL QUANTITY LIMITS: SAME AS FEDERAL

Louisiana*
Department of Environmental Quality
Hazardous Waste Division

P.O. Box 44307
Baton Rouge, LA 70804
(504) 342-1227
SMALL QUANTITY LIMITS: GENERATE UP TO 100 KG/MONTH, ACCUMULATE UP TO 1000 KG. ONCE AT THAT LIMIT, DISPOSE OF IN 90 DAYS. NO STORAGE BEYOND ONE YEAR UNDER ANY CIRCUMSTANCES.

Maine
Department of Environmental Protection
Bureau of Oil and Hazardous Materials Control
State House Station #17
Augusta, ME 04333
(207) 289-2951
SMALL QUANTITY LIMITS: NON-ACUTE WASTES 100 KG, ACUTE WASTES 1 KG, SAME LIMITS FOR SPILL MATERIAL

Maryland*
Department of Health and Mental Hygiene
Maryland Waste Management Administration
Office of Environmental Programs
201 W. Preston Street, Room A3
Baltimore, MD 21201
(301) 225-5709
SMALL QUANTITY LIMITS: 1 KG ACUTE WASTE, ALL ACUTE CONTAINERS OVER 20 LITERS, 10 KG OF INNER LINERS CONTAMINATED WITH ACUTE WASTE, 100 KG OF CONTAMINATED SOIL, 1000 KG OF NON-ACUTE WASTE

Massachusetts*
Department of Environmental Quality Engineering
Division of Solid and Hazardous Waste
One Winter Street, 5th floor
Boston, MA 02108
(617) 292-5589, (617) 292-5851
SMALL QUANTITY LIMITS: 1 KG ACUTE WASTE, 100 KG NON-ACUTE WASTE

Michigan
Michigan Department of Natural Resources
Hazardous Waste Division
Waste Evaluation Unit
Box 30028
Lansing, MI 48909
(517) 373-2730
SMALL QUANTITY LIMITS: SAME AS FEDERAL

Minnesota*
Pollution Control Agency
Solid and Hazardous Waste Division
1935 W. County Road, B-2
Roseville, MN 55113
(612) 296-7282
SMALL QUANTITY LIMITS: NO EXEMPTIONS FOR SMALL QUANTITY GENERATORS

Mississippi*
Department of Natural Resources
Division of Solid and Hazardous Waste Management
P.O. Box 10385
Jackson, MS 39209
(601) 961-5062
SMALL QUANTITY LIMITS: SAME AS FEDERAL

Missouri*
Department of Natural Resources
Waste Management Program
P.O. Box 176
Jefferson City, MO 65102
(314) 751-3176 Hotline: (800) 334-6946
SMALL QUANTITY LIMITS: SAME AS FEDERAL

Montana*
Department of Health and Environmental Sciences
Solid and Hazardous Waste Bureau
Cogswell Building, Room B-201
Helena, MT 59620
(406) 444-2821
SMALL QUANTITY LIMITS: SAME AS FEDERAL

Nebraska*
Department of Environmental Control
Hazardous Waste Management Section
P.O. Box 94877
State House Station
Lincoln, NE 68509
(402) 471-2186
SMALL QUANTITY LIMITS: SAME AS FEDERAL

Nevada*
Division of Environmental Protection
Waste Management Program
Capitol Complex
Carson City, NV 89710
(702) 885-4670
SMALL QUANTITY LIMITS: SAME AS FEDERAL

New Hampshire*
Department of Health and Human Services
Office of Waste Management
Health and Welfare Building
Hazen Drive
Concord, NH 03301-6527
(603) 271-4608
SMALL QUANTITY LIMITS: 1 KG ACUTE, 100 KG SPILL RESIDUE, 100 KG NONACUTE WASTE

New Jersey*
Department of Environmental Protection
Division of Waste Management
32 E. Hanover Street, CN-028
Trenton, NJ 08625
(609) 292-8341
SMALL QUANTITY LIMITS: 1 KG ACUTE, 100 KG SPILL RESIDUE, 100 KG NONACUTE WASTE

New Mexico*
Environmental Improvement Division
Ground Water and Hazardous Waste Bureau
Hazardous Waste Section
P.O. Box 968
Santa Fe, NM 87504-0968
(505) 827-2922
SMALL QUANTITY LIMITS: SAME AS FEDERAL

New York*
Department of Environmental Conservation
Bureau of Hazardous Waste Operations
50 Wolf Road, Room 209
Albany, NY 12233
(518) 457-0530 SQG HOTLINE: (800) 631-0666

SMALL QUANTITY LIMITS: SAME AS FEDERAL

North Carolina*
Department of Human Resources
Solid and Hazardous Waste Management Branch
P.O. Box 2091
Raleigh, NC 27602
(919) 733-2178
SMALL QUANTITY LIMITS: SAME AS FEDERAL. CONDITIONALLY EXEMPT GENERATORS CANNOT SEND HAZARDOUS WASTE TO MUNICIPAL LANDFILLS.

North Dakota*
Department of Health
Division of Hazardous Waste Management and Special Studies
1200 Missouri Avenue
Bismarck, ND 58502–5520
(701) 224-2366
SMALL QUANTITY LIMITS: SAME AS FEDERAL

Ohio
Ohio EPA
Division of Solid and Hazardous Waste Management
361 E. Broad Street
Columbus, OH 43266–0558
(614) 466-7220
SMALL QUANTITY LIMITS: SAME AS FEDERAL

Oklahoma*
Waste Management Service
Oklahoma State Department of Health
P.O. Box 53551
Oklahoma City, OK 73152
(405) 271-5338

SMALL QUANTITY LIMITS: SAME AS FEDERAL

Oregon*
Hazardous and Solid Waste Division
P.O. Box 1760
Portland, OR 97207
(503) 229-6534
SMALL QUANTITY LIMITS: SAME AS FEDERAL

Pennsylvania*
Department of Environmental Resources
Bureau of Waste Management
P.O. Box 2063
Harrisburg, PA 17120
(717) 787-6239
SMALL QUANTITY LIMITS: SAME AS FEDERAL

Puerto Rico
Environmental Quality Board
P.O. Box 11488
Santurce, PR 00910
(809) 723-8184
SMALL QUANTITY LIMITS: SAME AS FEDERAL

Rhode Island*
Department of Environmental Management
Division of Air and Hazardous Materials
Room 204, Cannon Building
75 Davis Street
Providence, RI 02908
SMALL QUANTITY LIMITS: NO EXEMPTIONS FOR SMALL QUANTITY GENERATORS

South Carolina*
Department of Health and Environmental Control
Bureau of Solid and Hazardous Waste Management
2600 Bull Street
Columbia, SC 29201
(803) 734-5200
SMALL QUANTITY LIMITS: LESS THAN 100 KG EXEMPT, OVER 100 KG REGULATED AS LARGE GENERATOR

South Dakota*
Department of Water and Natural Resources
Office of Air Quality and Solid Waste
Foss Building, Room 217
Pierre, SD 57501
(605) 773-3153
SMALL QUANTITY LIMITS: SAME AS FEDERAL

Tennessee*
Division of Solid Waste Management
Tennessee Department of Public Health
701 Broadway
Nashville, TN 37219–5403
(615) 741-3424
SMALL QUANTITY LIMITS: SAME AS FEDERAL

Texas*
Texas Water Commission
Hazardous and Solid Waste Division
Program Support Section
1700 N. Congress
Austin, TX 78711
(512) 463-7761
SMALL QUANTITY LIMITS: SAME AS FEDERAL (PROPOSED RULE: NO CONDITIONAL EXEMPTIONS)

Utah*
Department of Health
Bureau of Solid and Hazardous Waste Management
P.O. Box 16700
Salt Lake City, UT 84116–0700
(801) 538-6170
SMALL QUANTITY LIMITS: SAME AS FEDERAL

Vermont*
Agency of Environmental Conservation
103 S. Main Street
Waterbury, VT 05676
(802) 244-8702
SMALL QUANTITY LIMITS: 1 KG ACUTE WASTE, 100 KG NON-ACUTE WASTE; SAME FOR SPILL RESIDUES

Virginia*
Department of Health
Division of Solid and Hazardous Waste Management
Monroe Building, 11th floor
101 N. 14th Street
Richmond, VA 23219
(804) 225-2667 Hotline: (800) 552-2075
SMALL QUANTITY LIMITS: SAME AS FEDERAL

Washington*
Department of Ecology
Solid and Hazardous Waste Program
Mail Stop PV-11
Olympia, WA 98504–8711
(206) 459-6322 or (800) 633-7585 in state

SMALL QUANTITY LIMITS: SAME AS FEDERAL

West Virginia
Division of Water Resources
Solid and Hazardous Waste/Ground Water Branch
1201 Greenbrier Street
Charleston, WV 25311
SMALL QUANTITY LIMITS: SAME AS FEDERAL

Wisconsin*
Department of Natural Resources
Bureau of Solid Waste Management
P.O. Box 7921
Madison, WI 53707
(608) 266-1327
SMALL QUANTITY LIMITS: SAME AS FEDERAL

Wyoming
Department of Environmental Quality
Solid Waste Management Program
122 W. 25th Street
Cheyenne, WY 82002
(307) 777-7752
SMALL QUANTITY LIMITS: SAME AS FEDERAL

CHAPTER 2

What EPA Does Not Understand about Academic Laboratories

Peter A. Reinhardt

INTRODUCTION

Many school laboratories have not encountered problems with the U.S. Environmental Protection Agency's (EPA) hazardous waste rules; the rules are not widely understood by academic waste generators, nor are they actively enforced for the great majority. Schools gain an awareness of the hazardous waste regulations (in place since 1980) when they attempt to dispose of laboratory chemicals or if they are inspected by EPA or a state regulatory authority. Thus, larger schools—with larger waste streams that are more likely to be inspected—have had more problems with EPA's hazardous waste regulations. New rules for small quantity generators of hazardous waste, promulgated in 1985, have brought most academic laboratories under regulation. Enforcement has become stricter, too. In one case last year, a well-intentioned query to a regional office brought on an inspection and a fine.

Like all environmental regulations, hazardous waste rules are complex, voluminous, and difficult to understand. Unlike most environmental regulations, rules for the generation, transport, storage, treatment, and disposal of hazardous waste must be understood by teachers, researchers, building managers and chemical stockroom clerks—whomever is given the responsibility for regulatory compliance at an institution.

EPA regulations were directed at industrial generators of chemical wastes. They *were* also meant, however, to apply to academic laboratories, even though EPA estimated in 1979 that colleges and universities generate less than 1% of the nation's hazardous waste (Office of Solid Waste and Emergency Response, EPA, 1979). Likewise, EPA's inspection, permitting, and enforcement focus has been on large industrial hazardous waste generators. A veteran EPA inspector, when observing University of Wisconsin—Madison's shelves of vials and jars containing hazardous waste, admitted that he had never been to a hazardous waste storage facility with containers smaller than 55-gal drums.

EPA does not understand academic laboratories. They do not understand how chemicals are purchased, used, stored, and discarded in the process of research and teaching. They do not understand what makes a chemical a waste, or what makes a waste "hazardous" in the generic sense of the word. Therefore, it is not surprising that EPA has difficulty applying hazardous waste regulations to an academic setting. As the American Chemical Society (ACS) Task Force on the Resource Conservation and Recovery Act (RCRA, the act that authorizes EPA to regulate hazardous waste) stated when commenting on EPA's draft report to Congress, "Waste-producing operations in schools, such as those in the laboratory, are not scaled-down versions of industrial processes; and as such, the current RCRA rules are not appropriate or effective regulatory mechanisms" (ACS, 1987).

Common Complaints and Problems

The two most popular complaints about EPA's regulations put forward by laboratory generators have been that the record keeping required for generation, analysis, and transportation is too complex and detailed for such a varied waste stream, and that compliance with the storage time limits is not possible given the difficulty in disposing of some wastes and the inordinately high cost of small shipments. The two most common general problems with EPA's compliance programs, as cited by educational institutions, is that EPA or state regulatory authorities apply hazardous waste rules inconsistently in an academic setting and that a regulator's interpretations often do not conform to the regulatory intent of RCRA. Inconsistent application of the rules occurs when regulators find situations that only remotely resemble those in industry. For example, schools within the same state and EPA region have been assigned I.D. numbers using different criteria for an "individual generation site." Also, when regulatory authorities prevent the treatment of chemicals in laboratories they contradict Congress' objective of waste minimization.

A brief discussion of the history and progress in suggesting regulatory reform of the rules for hazardous waste management in academic laboratories follows.

PROPOSED REGULATORY REFORM

In 1983 the National Research Council (NRC) recommended the following regulatory changes to EPA in their book, *Prudent Practices for the Disposal of Chemicals from Laboratories*:

1. Encourage consistency and integration among federal, state, and local regulations.
2. Reduce the detail required for waste characterization, reporting, shipping, and disposal by describing wastes according to seven classes.
3. Create a uniform manifest for lab packs.
4. Simplify record keeping by using aggregate units.

5. Continue to regard laboratory hazard reduction methods as unregulated.
6. Make special provisions for pre-tested, small-scale incinerators.

With respect to consistency with local regulations, EPA's role according to law is to set the minimum standards; local and state regulators are free to adopt more stringent standards. This is entirely appropriate within our political system. Items 2, 3, and 4 address the very burdensome record-keeping requirements: keeping a waste inventory, obtaining analysis information, and completing detailed manifests according to DOT regulations. Item 5 refers to a very critical exemption for bench-top chemical and physical treatment of hazardous waste in laboratories. Contrary to what is indicated, the exemption of in-lab chemical treatment processes is not universally accepted by regulatory authorities. As far as I can determine, EPA does not have a policy on this. EPA does exempt "Totally Enclosed Treatment Facilities" from hazardous waste facility requirements, and these activities may be considered as such.

In a 1985 *Environmental Science and Technology* article, Peter C. Ashbrook and I noted these needed changes in the regulations. The EPA should:

1. Establish hazard assessment procedures for chemicals not specifically regulated by EPA. Many toxic chemicals are not defined as hazardous in EPA regulations.
2. Exempt small or *de minimis* quantities or concentrations of hazardous waste.
3. Establish standard tests for the identification of unknowns. Analysis of unknowns (a fixture of laboratory wastestreams) should be done by an EPA approved method that can be done at a reasonable cost and is commonly accepted by disposal companies.
4. Raise the 1 kg limit for acute hazardous waste. The 90 day storage limit contingency planning and training requirement for acute hazardous waste is unreasonable.
5. Exempt schools from the large quantity generator requirements in the case of one-time generation. Even high schools generate 1 kg of cyanide salts occasionally; inspectors usually do not enforce the rules in this case.
6. Simplify record keeping by describing wastes according to their characteristics.
7. Offer incentives to companies that will service small quantity and remote generators. Businesses have not taken the risk to service small quantity laboratory generators. Safety Kleen, a commercial solvent recycling firm, has proven that the small quantity business can be profitable.
8. Simplify the permitting procedures for small, bench-top incinerators.
9. Research and establish standard methods for the destruction of potentially explosive materials. This is important for EPA and generators alike—to limit long-term liability by finding processes that reduce the potential for a release.

Mr. Ashbrook and I also raised these concerns with EPA directly. We wrote and met with Dr. Alan Corson (now retired) of the EPA Office of Solid Waste, who represented EPA on ACS's RCRA Task Force. In our correspondence with him (30 July 1986 and 8 October 1986), we noted the additional need to clarify the definition of what constitutes an acceptable hazardous waste "analysis" and to clear up problems with the definition of "on-site" and "contiguous" for the purposes of assigning I.D. numbers and determining when manifesting is required. In some cases, schools were required to routinely analyze chemicals in their original, labeled containers. Permits for commercial incinerators usually allow a wide variety of constituents, so the label or a material safety data sheet is sufficient analysis information.

EPA EFFORTS TO UNDERSTAND EDUCATIONAL INSTITUTIONS

EPA is trying to better understand academic laboratories. In 1983 Congress directed EPA to prepare a report on problems associated with the management of hazardous waste from educational institutions. The report was drafted by Tufts University and ICF, Inc., a consulting firm. The draft report was reviewed by ACS, the National Association of College and University Business Officers (NACUBO), the Department of Education, and other bodies representing educational institutions. EPA considered these comments and made some changes to the report, which was then published in April 1989 (National Technical Information Service Document-PB89–187–629).

The report addresses the problems of hazardous waste management faced by all educational institutions in the U.S.: 23,000 high schools, 3,000 vocational schools, and 3,000 colleges and universities. Only seven secondary schools were surveyed, however. EPA states that "the study was not intended to be a statistical survey but rather a series of case studies to identify problems" (Marcia E. Williams, EPA Office of Solid Waste, in a letter to Kathleen A. Ream, ACS Department of Government Relations and Science Policy, 8 January 1988). The draft report contained five regulatory reform recommendations (ICF, Inc., 1987). They suggested that EPA should:

- Ensure uniform federal, state and local regulations (taken from NRC No. 1, 1983).
- Simplify waste identification (NRC No. 2, 1983).
- Reduce the requirements for small incinerators (NRC No. 6, 1983).
- Exempt the transport of small quantities.
- Change the definition of "on-site."

The last two recommendations were to address Mr. Ashbrook's and my concerns regarding I.D. numbers and manifesting requirements.

Issues Raised by Commentators on the Draft Report

ACS's Task Force on RCRA commented extensively on the draft report. They pointed out that "by identifying educational institutions as a special class of SQGs [Small Quantity Generators], the Congress was suggesting that special regulations—different from those for other SQGs—might be needed"

(ACS, 1987). ACS recommendations that will be incorporated into EPA's Report include:

- Define academic laboratories in the regulations.
- Allow on-site storage of laboratory waste for more than 90 days without a TSD permit (e.g., up to one truckload or one year).
- Simplify lab pack waste identification requirements according to NRC's recommendations.
- Establish an independent office to assist academic waste generators.

By incorporating ACS's comments into the report, EPA has not agreed to make any regulatory changes. Instead, EPA is simply indicating that these changes are possible and are appropriate for inclusion in the report to Congress. It is up to us to prod Congress to enact regulatory change or ask EPA to do so.

In separate comments, the National Association of Colleges and University Business Officers (NACUBO) concurred with ACS that alternatives to TSD permits are needed to accumulate academic wastes (Mary Jane Calais, NACUBO, in a letter to Michael Petruska, EPA Office of Solid Waste, 30 October 1987. Note that NACUBO has since published the book, *Hazardous Waste Management at Educational Institutions*). Interestingly, NACUBO cautioned EPA about adopting a simplified record-keeping scheme, fearing that records might be incompatible with disposal company requirements. Also, NACUBO suggested that an enforcement amnesty program be conducted (including a publicity campaign by EPA) to encourage educational institutions to establish hazardous waste management programs.

Other Comments

Peter Ashbrook of the University of Illinois commented independently that the report hardly mentions the problem of wastes that are difficult to dispose of: explosives, reactives, 2,4,5-T, dioxin precursors, gas cylinders and unknowns. He also felt that the impact of the landfill ban should have been covered (Peter Ashbrook, in a letter to Filomena Chau, EPA Office of Solid Waste, 18 October 1987).

It is my feeling that the report should not attempt to estimate the waste generation rates at schools without programs to routinely collect hazardous waste because such waste is simply stored indefinitely or disposed of illegally. The report refers to the option of a school obtaining a permit for storage for more than 90 days. However, very few schools have the resources to overcome the requirements for permitting, siting, public relations, and satisfying a permit writer who does not understand academic laboratories and has the authority to ask endless questions.

It may also be noted that the report to Congress did not contain recommendations on tests for unknowns, treatment of explosives, servicing small quantity generators or what constitutes analysis.

MORE THOUGHTS ON REGULATORY REFORM

Simplified Characterization and Record Keeping

I am in agreement with NACUBO's suspicion about the success of requiring less detail. For liability purposes, commercial facilities will continue to want accurate descriptions of a lab pack's contents. For incineration they want to know, for example, the Btu, chlorine, and metal content of the lab pack, which is only obtainable by adding the values of each component. They also want to know what waste belongs to whom, so that in the event of a release they can share cleanup costs. In addition, schools are not packing their own lab packs (and doing the commensurate paperwork) anymore; most commercial disposal firms have assumed that task. An up-to-date chemical inventory may be burdensome, but I feel it is necessary for chemical safety. By using a personal computer this information can be used to prepare manifests. DOT rules have required detailed shipping papers for transport of all chemicals since the '60s.

I.D. Numbers, Sites, and Manifests

This issue is explored extensively in a *NACUA Law Digest* article (1980) by Fred K. Heinrich, an attorney from the University of Illinois. In it, Mr. Heinrich discusses the definitions of "generator," "on-site," and "individual generation site." Mr. Heinrich explains that EPA's goal is to control the generation, transport, storage, treatment, and disposal of hazardous waste. This is the purpose of the "cradle-to-grave" manifest system. It is EPA's policy, however, to consider each block a "site" and an individual generator; in most cases this means a school that is a large quantity generator becomes instead many small quantity generators. EPA's policy results in less control of hazardous waste due to the small quantity generator exemptions. Also, manifesting between sites and site reports create a paperwork nightmare for both EPA and the school.

It should be noted that chemical plants and industrial plants have sites intersected with public thoroughfares too; the difference is that at many educational institutions, there are more roads dividing the site into smaller plots.

OPTIONS FOR REGULATORY REFORM

There are four options to change the way EPA's hazardous waste regulations affect laboratories:

1. *Amend RCRA*. This would take an act of Congress and considerable lobbying. Organizations representing educational institutions have more important issues to attend to with their limited resources.
2. *Negotiate to Amend the Regulations.* The best way to do this would be by means of an Advisory Committee that includes representatives from educa-

tional institutions and other generators of laboratory waste (For an example of how this works, see the *Federal Register*, 51 FR 25739–25742).

3. *Issue a Guidance Manual* for inspectors, enforcement personnel, and permit writers that defines specific policies to be followed for colleges and universities. Ideally, this guidance would be negotiated by means of an advisory committee.
4. *Issue Policy Statements* to be made available through the Office of Solid Waste's Directives System. (According to EPA, each region's Policy Directives Coordinator has a list of all RCRA-related policy, guidance, and memoranda, and where they can be obtained.) Again, it would be best if these policies were negotiated by means of an advisory committee. There is a document, "RCRA Permit Policy Compendium: Summaries," that includes EPA's interpretation of the regulations with respect to educational institutions (EPA 9451.02[83]). Some of these policies are answers to questions received on the EPA Hotline.

Most EPA guidance manuals are written by consultants, and the EPA Hotline is operated by a consultant. The manual and Hotline answers are reviewed and approved of by EPA. It is my opinion, however, that in many cases it is consultants (rather than EPA) that establish policy—and some consultants have a vested interest in the interpretation of regulations. (Mitre Corporation's guidance manual written for EPA on permitting incinerators defined an insignificant concentration of hazardous constituents as 100 ppm. At greater concentrations, Mitre will perform a trial burn of an incinerator at a cost of $20,000 to $100,000.) Also, it is very difficult to reverse an interpretation after EPA's initial approval; a casual answer to a seemingly trivial question can have very significant policy implications. The advisory committee approach appears to allow a broader perspective into the regulatory/policy-making process.

WHAT CAN BE DONE

First and foremost, laboratories and schools should take full advantage of the provisions of EPA's regulations that favor laboratory waste. These provisions should be exercised when it is prudent. At all times *schools should manage all hazardous materials (whether or not they are wastes) in a manner that protects human health and the environment*. Specific provisions that favor laboratories include:

- A waste is not a waste until it is given that designation. Until all reasonable alternatives have been explored, the material is a surplus chemical.
- Non-commercial transport of surplus chemicals is exempt from EPA and DOT regulations. Storage of surplus chemicals is exempt from EPA regulations.
- Recycling is a regulated activity; redistribution of surplus chemicals is not.
- Many chemical wastes are not hazardous waste under EPA regulations. U and P lists only pertain to discarded commercial chemical products and

spills; wastes contaminated from the use of U and P wastes are not regulated. *Many unregulated chemical wastes should be handled as hazardous waste*, but record-keeping requirements and storage time limits do not need to be complied with.

Coordinated Efforts for Regulatory Reform

The EPA report to Congress provides an excellent opportunity to gain congressional attention for amending RCRA or asking EPA to issue alternative rule making. I suggest that interested parties pursue these changes through ACS, NACUBO, the American Council on Education, or other organizations that represent educational institutions.

BIBLIOGRAPHY

Ashbrook, P. C., and P. A. Reinhardt. *J. Environ. Sci. Technol.* 19(12):1150–1155 (1985).

Heinrich, F. K. "To Be Or Not To Be, A Single Contiguous Site: The U.S. EPA Carves Up College Campuses," *NACUA College Law Digest* 18(5):104–114 (1980).

ICF, Inc. *Problems Associated with Management of Hazardous Waste from Educational Institutions: Draft Report* (Washington, DC: U.S. Environmental Protection Agency, September 1987).

National Research Council. *Prudent Practices for Disposal of Chemicals from Laboratories* (Washington, DC: National Academy Press, 1983).

Office of Solid Waste and Emergency Response. *Technical Environmental Impacts of Various Approaches for Regulating Small Volume Hazardous Waste Generators*, Vol. I (Washington, DC: U.S. Environmental Protection Agency, 1979).

"RCRA Permit Policy Compendium: Summaries," U.S. Environmental Protection Agency-Document 9451.02(83) (19 October 1983).

"Report to Congress: Management of Hazardous Wastes from Educational Institutions," U.S. Environmental Protection Agency, National Technical Information Service-Document PB89–187–629 (April 1989).

Task Force on RCRA. *Peer Review of the USEPA's Draft Report to Congress: Problems Associated with Management of Hazardous Waste from Educational Institutions* (Washington, DC: American Chemical Society, 23 October 1987).

CHAPTER 3

Unwanted Chemicals in Secondary School Laboratories

Thomas Kelley

INTRODUCTION

The purpose of this chapter is to discuss the latest efforts of the Laboratory Safety Workshop, a national center for training and information located at Curry College in Massachusetts, to remove unwanted chemicals from secondary school laboratories. The Laboratory Safety Workshop generates numerous publications, including a newsletter, *Speaking of Safety* and a brochure, *Laboratory Safety Guidelines*. Among its many other services, it runs a free audio-visual lending library and seminars on laboratory safety for science teachers. Those who have attended the Center's seminars rate the disposal of unwanted chemicals as their number one problem.

Dr. Kaufman, the Director of the Center, and I got together in September 1986. He invited me to coordinate a project that involved a survey of 800 Massachusetts secondary schools to determine if there existed a need for a general cleanout. Questionnaires were mailed to science chairpersons asking them to list their unwanted chemicals. We also asked them several related questions on health and safety.

This chapter will describe our findings and discuss the direction in which we are heading.

FINDINGS OF THE LABORATORY SAFETY WORKSHOP SURVEY IN MASSACHUSETTS

Secondary School Disposal of Unwanted Chemicals

There were two separate mailings in the fall of 1986. The total response rate was a little better than 12%. Out of nearly 100 schools participating in our survey, 44% expected to participate in "Operation CleanSweep." This is a program sponsored by the Massachusetts Department of Environmental Management to rid towns of their household hazardous wastes. Schools could take part in this program. Thirty-six percent were either unsure or unaware that such a program existed. Unfortunately, "Operation CleanSweep" ceased in July 1987 because of lack of funds.

Unwanted chemicals were measured in terms of pounds and gallons. The survey responses revealed approximately 8700 lb and 500 gal in the 100 participating Massachusetts secondary schools.

We wanted to find out the last time schools had called a disposal contractor to have their chemicals removed. Thirty-seven percent said that their chemicals had been removed by a contractor between 1985–1987. The majority, 48%, said chemicals had never been removed. Several wanted a reimbursement if funds were appropriated for a one-time general cleanout. Dr. Kaufman reminded me that this was Massachusetts and not Fantasy Island.

We also wanted to learn the reasons why teachers had not bothered to call a disposal contractor in the past. Twenty-five percent responded that cost was the major factor keeping them from disposing of their unwanted chemicals. Twenty percent said lack of knowledge was the most important reason. Nineteen percent cited other disposal problems such as not knowing who to call or lack of available facilities. Nine percent mentioned safety concerns for the environment.

Secondary School Laboratory Accidents—A Related Issue

There have been many tragic examples of school accidents recorded. In one example, four children at a Washington, DC elementary school laboratory were severely burned after combustible powders they and their classmates were combining to create "sparklers," blew up. In another example, two fifteen-year-olds at Attleboro High School were seriously burned as the result of a fire, probably caused by the ignition of fumes from duplicating fluid. Unfortunately, however, many elementary and secondary school accidents go unreported. In order to learn more about accidents, the Laboratory Safety Workshop also surveyed the number and types of laboratory accidents that occurred in Massachusetts secondary schools.

The responses showed a total of 87 laboratory accidents in the 1986 academic year. Fifteen were deemed serious. Most of the accidents reported were the result of broken glass. The more serious ones ranged from electrical shocks to burns from hot equipment. There were also a few cases of facial burns from acid. This is an accident rate 50% higher than the chemical industry and 35 times worse than DuPont. Similar results have been obtained in the Center's surveys of college and university chemistry departments.

We also counted the number of safety coordinators in science departments. Seventy-two percent of school science departments did not have a safety coordinator. Twenty percent did have a person assigned to that role. When we looked at schools as a whole, we found that 86% did not have a safety coordinator. Eight percent of the schools surveyed had a person assigned to that role.

We feel that every laboratory should be equipped with a written laboratory safety program for both teacher and student to follow. To address this area, we asked each science chairperson whether his or her school had such a written program. Sixty-three percent of those responding stated that no written laboratory safety program existed. Thirty percent responded that they did indeed have a written laboratory safety program in place.

PRESENT AND FUTURE ACTION FOR SECONDARY SCHOOL CHEMICAL DISPOSAL IN MASSACHUSETTS

Present Legislation

In Massachusetts, the legislature passed the Solid Waste Act of 1987. Under Chapter 584 of this Act, high school hazardous waste generated in science classes is subject to proper disposal under rules and regulations promulgated by the Department of Environmental Quality Engineering (DEQE).

In addition, the legislature is considering H.1749, a bill designed to promote the reduced use of toxic and hazardous substances in the Commonwealth. The bill is currently under consideration in the Natural Resources Committee. Following issuance of a committee report, it is anticipated that the proposal will be forwarded to the full House for deliberation and action.

The Need for More Legislation

Although this is a good beginning, I believe a bill has to be filed that will empower the DEQE to perform a one-time general cleanout.

This program could be modeled after the successful one operating out of the University of Minnesota where unwanted chemicals are collected in the fall and spring of each year, brought to certain locations in the state, and transported for disposal.

Clean Harbors, Inc.

In Massachusetts one company is helping to meet the need for unwanted chemical collection. Clean Harbors, Inc., the largest waste disposal company in New England, recently sent invitations to all superintendents, principals, and science coordinators of secondary schools in Massachusetts to participate in a $100,000 lottery for free chemical collection services. Eighty-one applications were received before the 1 December 1987 deadline. Each applicant community was represented in the drawing. Applicants on the list were asked to submit final inventories of chemicals that required removal by 1 January 1988. Clean Harbors began collections on 4 January 1988. Seventeen schools were serviced in ten communities before the funds were depleted. The chemicals disposed ranged from common flammable solvents to reactives and explosives.

Massachusetts Department of Environmental Management

The Massachusetts Department of Environmental Management (DEM) has set different guidelines in the collection of hazardous chemicals in secondary schools. Under the new system, these schools are now considered regulated generators of hazardous waste. This stipulation makes them ineligible for grant money.

DEM is running workshops on health and safety at different locations throughout the state. These workshops are intended to help school administrators and teachers promote health and safety in their classrooms.

The Chemical Management System of the Laboratory Safety Workshop

I would like to see schools and colleges set up a Chemical Management System as proposed by the Laboratory Safety Workshop. The major components of this system would include purchasing, inventory, storage, use, and disposal of chemicals. Such a system is described in Chapter 6.

In this system, the stockroom coordinator would indicate the date of the chemical purchase. Old chemicals would be discarded after their expiration date. In addition, the chemicals would be segregated according to hazard class. Furthermore, he would make sure a material data safety sheet had been supplied by the vendor. The appropriate data sheets would be provided to each science teacher. These sheets give a wealth of information on the chemical and physical properties of the material as well as emergency guidelines to follow in the event of a chemical accident. Teachers would apply the same procedures once the chemical was brought into their laboratory. The teacher would have the responsibility of strictly enforcing this segregation with his students.

The Chemical Management System would include the setting up of a segregation system for waste generated from chemical experiments. This would begin in the laboratories where students are working. Segregation of laboratory wastes helps to encourage recycling and reduces the likelihood of mixing reactive incompatibles.

The adoption of an inventory system would allow school officials to implement other important safety measures. One of these would be the posting of National Fire Protection Association diamond labels on the outside of each laboratory where chemicals are stored. These labels designate potential health, flammability, and reactivity hazards for emergency personnel (fire fighters) entering a room. The diamond labels would be posted following the completion of a chemical inventory. A record would be kept of the labels and their corresponding laboratory. These numbers would be checked annually for accuracy after a review of chemical stock.

An inventory system would also allow the substitution of non-hazardous or less hazardous chemicals for hazardous ones. An examination of the chemicals used could be done and substitutes recommended that produce the same result but pose less of a threat to human health. The Minnesota Department of Education has published a helpful booklet in this area, *New Chemicals for Old*, suggesting several such substitutions.

Kell's Chemical Inventory Service

I started Kell's Chemical Inventory Service to promote health and safety in secondary schools through inventory planning and emergency response. This service implements some of the points I have just mentioned.

Chemswap

The Laboratory Safety Workshop is developing a strategy to establish a chemical exchange program between secondary schools. This program will be known as CHEMSWAP. We hope to help schools reduce the cost of purchasing new chemicals by exchanging unwanted ones. This will also cut down in disposal costs. Sixty-five percent of schools who responded in our survey would like to participate in the CHEMSWAP program. Twenty-one percent were uncertain. Fourteen percent stated that they were not interested. Any group of neighboring institutions can benefit from such a program. Sharing lists of unwanted chemicals saves money.

It is our hope that grant money will be available in the upcoming months that will allow the Laboratory Safety Workshop to start the program. Schools that have already indicated the chemicals they do not want will have the opportunity to specify the chemicals they do need.

In addition, I would like to see schools begin budgeting for the cleanout of unwanted chemicals on a regular basis. Each school would designate a portion of their budget for health and safety. The money would also allow school administrators to address their other health and safety needs.

CONCLUSION

Last but not least, I hope Dr. Kaufman continues his fine work in instructing science teachers in laboratory safety through his Laboratory Safety Workshop at Curry College.

This brief chapter has been designed to give an idea of our efforts to assist schools in dealing with the problem of unwanted chemicals. One important role interested individuals can play in the process is to inform state senators and representatives of their views on the need for state action.

The Laboratory Safety Workshop remains ready to assist with laboratory safety questions or problems.

SECTION II

Academic Waste Disposal Programs

CHAPTER 4

Surviving a Disposal Crisis in the Small Academic Laboratory

Walter J. Warner, Jr.

INTRODUCTION

This chapter documents the problem-solving process we use for hazardous/unwanted waste disposal in our small private secondary school science department. All of the solutions to this problem have been successfully used in other settings. To my knowledge, on-site disposal has not been used by private secondary schools in the state of Connecticut to date. I will attempt to chronicle the particular approach used, focusing on some specific implementation problems encountered in the small private school setting. Attempts to solve these specific problems will be discussed. My hope is that this chapter will (1) serve as an initial resource for teachers who inherit a waste disposal problem (and do not have access to a simple solution); and (2) serve as a catalyst for states to address the problem of hazardous waste management in private academic institutions with limited funds to spend on such waste disposal.

Development of our plan initially involved outlining an overall problem-solving strategy:

- I. Organizational Phase
  - a) teacher education/skill development
  - b) enlisting help
  - c) taking a thorough inventory
- II. Initial Disposal of Accumulated Waste
  - a) trading/giving chemicals away
  - b) state/local pickup
  - c) on-site disposal
  - d) safe storage of chemicals not expeditiously disposed of
- III. On-going Waste Management
  - a) restructuring laboratory curricula
  - b) on-site disposal
  - c) safe storage
  - d) prudent purchasing practices

I note here that no chemicals should be disposed of in any way until the teacher has gained sufficient knowledge and confidence to work with them (NEVER WORK ALONE!). This is a subjective judgment on the teacher's part, but I strongly suggest following these *minimal guidelines* before tackling the problem:

1. Read background material including *Prudent Practices for Disposal of Chemicals from Laboratories*, National Research Council (NRC); *Safety in Academic Chemistry Laboratories*, American Chemical Society (ACS); and *Chemical Catalogue/ Reference Manual*, Flinn Scientific (see Appendix I).
2. Attend a laboratory safety workshop (see Appendix II).
3. Know specific state and local hazardous waste management regulations.
4. Acquire specific skills, including basic first aid, specific first aid for chemical accidents, basic laboratory skills, and proper cleanup of chemical spills. (See Appendix I for a list of appropriate references.)

Enlisting Help

See Appendix II for a list of contact agencies/people.

In a small school both departmental and administrative support is vital. Getting the financial support needed at our school was facilitated by involving the plant manager and insurance adjustor early in the problem-solving process. This was accomplished during a routine plant inspection and in a follow-up meeting to discuss specific suggestions to comply with insurance regulations.

The first outside contact person I suggest engaging is the science coordinator at the State Department of Education. Even if (as was the case in my situation) this person cannot directly help with the problem, he/she is an excellent resource for identifying those who can help.

Regardless of the disposal methods chosen, finding a reliable person at the State Department of Environmental Protection (DEP) is crucial. This is important to facilitate acquisition of knowledge of state and local regulations. This person can also act as liaison if permits are needed for on-site disposal, if wastes are to be transported via the mail, or if a reputable local contractor is needed. In some cases, it may be helpful to enlist a contact person at the United States Environmental Protection Agency (EPA), particularly if state and local authorities are reluctant to act without specific EPA direction.

Before deciding which of the several disposal options or combination of options best suited my resources and funds, I defined the scope and nature of the problem by taking a thorough inventory. The Flinn *Reference Manual* states:

> Such an inventory would serve many valuable purposes such as (but not limited to): . . .
>
> • To rid the premises of excess/unused chemical substances . . .

- To identify substances (severe toxins, carcinogens, etc.) that should not be found on school premises and rid the premises of these materials. . . .

All appropriate safety precautions should be strictly adhered to during this endeavor! The Flinn *Reference Manual* presents basic guidelines for this process. (See also Fischer's article in Appendix III.)

Upon completion of this task, careful consideration was given to develop a comprehensive and cost-efficient plan for disposing of unwanted/hazardous chemicals. Factors considered included budgetary restrictions, my confidence in handling these chemicals, and local and state waste management regulations.

Due to the limited financial resources available to me, I attempted a combination of solutions to the problem. First, I submitted an excess chemical removal quotation form to a national firm, Chem Services, Inc., for a price quotation, and I contacted local firms to get cost estimates of local disposal. Meanwhile, the following methods were used to reduce our inventory as much as possible to minimize the eventual cost of having such a contractor remove these chemicals:

1. We contacted local private schools and colleges offering to give away or trade away unwanted chemicals for chemicals we could use.
2. We utilized the town/state sponsored community household chemicals.
3. We neutralized and safely stored less hazardous by-products.
4. We used a properly licensed waste disposal contractor.

This phase is not complete, although we are making some progress.

ON-GOING WASTE MANAGEMENT

Prevention of a future crisis involves developing responsible, yet cost-efficient, chemical purchasing procedures. For schools with the appropriate student body and flexibility to restructure their laboratory curricula, a list of references to help initiate this process is found in Appendix IV. I helped make this a departmental project with the focus on introducing a number of interesting laboratory experiments in introductory biology and chemistry classes, requiring mostly chemicals from the local grocery or hardware store. Interesting experiments that involve the use of hazardous chemicals or dangerous reactions have been converted to demonstration activities to minimize the quantities of these chemicals kept on hand. An effort has been made with both the laboratory activities and demonstrations to minimize or eliminate reactions producing hazardous waste products.

Safe storage of small quantities of hazardous waste generated by students and teachers is vital, both to insure a safe laboratory environment, and to promote a responsible, conscientious attitude and approach to hazardous waste management in our students. ACS's *Safety in Academic Chemistry Laboratories* contains a helpful section entitled "Lab Bench Cleanup and Disposal of Chemicals," geared for students. The Flinn *Reference Manual* contains some helpful suggestions on safe longer term storage of unwanted chemicals in the section on taking a chemical inventory.

Hiring an outside contractor was a last resort for us due to the high cost. Nevertheless, it became necessary to use/pursue this option, since we had exhausted our other options and still had some unwanted hazardous chemicals. I recommend contacting state DEPs to obtain lists of licensed contractors in their areas.

On-site disposal of small quantities of unwanted chemicals is a difficult option in Connecticut if (as in my situation) the institution does not meet the minimum criteria for classification as a small quantity hazardous waste generator. Progress is being made in Connecticut toward making this a viable option for institutions that generate less than 100 kg/mo of hazardous waste. For a more detailed discussion of this option see the specific problems and solutions sections that follow.

SPECIFIC PROBLEMS ENCOUNTERED

1. Finding time to educate myself and do the work
2. Getting financial support for my time
3. Getting colleagues at other schools to help
4. Establishing effective and reliable contact people at the local, state, and federal levels
5. Complying with federal, state, and local regulations—especially with the on-site disposal option

SOLUTIONS

1. Vacations and the summer are the most reasonable times for me to work. With the myriad of duties and the hectic pace of the school year, it is difficult at best to accomplish much then.
2. I got enough financial support to undertake this project by a) convincing the school head I could actually save him money on insurance costs; b) incorporating the work into my Master's study, thereby allowing me to work in my lab while being financially supported by my institution; and c) agreeing to present my work here, thereby enlisting further financial aid from the school's conference and professional development budget.
3. Encouraging colleagues at my school to get involved was easy, but getting colleagues at other private schools to trade chemicals or take our unwanted chemicals (even if they use those chemicals in their laboratories) has been difficult. We sent out letters to other schools explaining our situation and asking for help. I also regularly contacted colleagues at local science teacher association meetings. This has been a moderately successful endeavor to date, although we reduced our inventory by only about 5% using this method.
4. Establishing effective and reliable contact people in the Connecticut DEP was

initially a frustrating problem. I attribute the success I have had to patience and persistence, rather than to skill. Making a pest of myself seemed to be the only solution. Announcing that I was writing on this subject seemed to get the DEP's attention. I needed only minimal contact with the federal EPA, so this was not a problem. Getting local officials to cooperate was not a problem for me either, although I was warned that potential problems with effective communication might prove to be very frustrating. I count myself lucky in this regard.

5. Initial attempts to comply with federal, state, and local hazardous waste regulations proved to be frustrating. Responses from officials at *each* level indicated we would have *their* approval as long as we did not violate the regulations of the *other* two levels. Achieving any resolution in this seemingly eternal merry-go-round took some doing.

   A memo dated 9 October 1986 issued by the Office of General Counsel of the federal EPA stated that for generators of less than 100 kg/mo of hazardous waste "all that is required is that the State have some mechanism for approving facilities that propose to manage the exempt waste. Since the underlying intent of the requirement is that the State assess the risks associated with particular facilities handling the exempt waste, any mechanism that the state chooses to accomplish this is, in our view acceptable under the regulations. Thus, we would not judge an exchange of letters to be an inappropriate way to achieve "registration" of a facility [The regulations do not define the term "registration."]." (see Appendix V)

This memo was vital in facilitating the process because the Connecticut State Regulations provide for this special registration of conditionally exempt small quantity generators in section 25-54cc(c-2ii), stating that small quantity generators may

> otherwise dispose of the hazardous waste under the direction and with the written approval of the Commissioner pursuant to section 19-524n of the General statutes and section 19-524-6(1) of the Regulation of State Agencies [see Appendix V].

Having established that institutions of our type may be eligible for this exemption, the DEP officials reviewed the proposed on-site disposal procedures presented in the Flinn *Reference Manual.* A cursory review indicated that at least some of the neutralization and disposal procedures do not produce hazardous waste in quantities that violate present Connecticut State Air and Water Regulations. The problem is that the air and water regulations are presently being rewritten, and it will be some time before the first draft of the proposed new regulations will be available for public scrutiny. I will participate in the review of the new regulations and have been assured by the DEP officials that they are anxious to help people in my situation solve their waste disposal problems safely, efficiently, and economically. They agree that on-site disposal is an efficient and economical alternative and are optimistic about the proposed methods, which are deemed safe and which do not violate the new air and water regulations. It is not our intention for DEP to endorse a particular set of disposal methods, but simply to make available alternatives that can be approved on a case by case basis.

CONCLUSION

My original goal when starting this project was simply to rid our science department of unwanted chemicals and make it a safer place to work and learn. I quickly learned that this was a complex problem requiring a detailed, carefully thought out strategy. I also became aware that an inexperienced and untrained person *could* begin to solve a problem of this nature, *and* that my attempts, however naive and uninformed, could help raise the level of awareness at the Connecticut DEP that some small, economically limited private institutions in Connecticut have a waste disposal problem to which the solution is not easily facilitated by convenient means.

Two more far-reaching goals have sprung from this work. These are 1) to help implement on-site disposal of small quantities of waste at conditionally exempt hazardous waste generating institutions as a viable disposal method; and 2) to publicize both the procedures for gaining a conditional exemption and the viability of safe, on-site disposal as means of reducing disposal costs.

The original project is nearing completion. We have reduced our inventory of unwanted chemicals from over 600 lb to under 150 lb. Some of the chemicals remaining are of a hazardous nature, but are safely and securely stored. We intend to use a licensed contractor to remove these chemicals. We hope to reduce our inventory of nonhazardous chemicals further by using on-site disposal methods once a state registration permit is granted.

APPENDIX I

Background Reading

American Red Cross. *Advanced First Aid and Emergency Care*, 2nd ed. (Garden City, NY: Doubleday & Co., Inc., 1979).

Chemical Catalog/Reference Manual (Batavia, IL: Flinn Scientific, Inc., 1988).

Lefevre, M. J. *First Aid Manual for Chemical Accidents* (for use with nonpharmaceutical chemicals) (New York: Van Nostrand Reinhold Company, 1989).

National Research Council. *Prudent Practices for Disposal of Chemicals from Laboratories* (Washington, DC: National Academy Press, 1983).

Office of Solid Waste. "Solving the Hazardous Waste Problem," U.S. EPA, EPA's RCRA program, EPA/530-sw-86-037 (November 1986).

Safety in Academic Chemistry Laboratories (Washington, DC: American Chemical Society, 1985).

APPENDIX II

Contact Agencies/People

National

Chem Services, Inc.
Van Nortwick Street
P.O. Box 426
Batavia, IL 60510
Telephone 312/879-6901

Flinn Scientific Company
P.O. Box 219
131 Flinn Street
Batavia, IL 60510
Telephone 312/879-6900

Laboratory Safety Workshop
Curry College
Milton, MA 02186
Telephone 617/333-0500 (x2220)

OSHA Assistance, Inspection,
and Training Materials
(See Appendix VI of *Safety in Academic Chemistry Laboratories*, ACS, for a
listing of regional OSHA offices.)

RCRA Hotline
Telephone 800/424-9346 (toll free)

U.S. Environmental Protection Agency
Office of Solid Waste (WH 562)
401 M. Street S.W.
Washington, DC 20460

Connecticut/New England

Connecticut State Department of Education
165 Capitol Avenue, Box 2219
Hartford, CT 06145
Contact: Sigmund Abeles
Telephone 203/566-4825

EPA Region I
Waste Management Division (HHA)
John F. Kennedy Building
Boston, MA 02203

Department of Environmental Protection
Hazardous Material Management Unit
State Office Building
165 Capitol Avenue
Hartford, CT 06106

Dews, George
DEP
Senior Sanitary Engineer
Hazardous Waste Management Section
Hazardous Materials Management Unit
122 Washington Street
Hartford, CT 06106
Telephone 203/566-5712 & 566-4869

DHHS NIOSH Regional Office
Region I
JFK Federal Building
Room 1401
Government Center
Boston, MA 02203
Telephone 617/223-3848

Giroux, Barry L.
DEP
Principal Sanitary Engineer
Hazardous Waste Management Section
Hazardous Materials Management Unit
122 Washington Street
Hartford, CT 06106
Telephone 203/566-5712 & 566-4869

APPENDIX III

Journal Articles and Books

American Chemical Society. *CHEMCOM Chemistry in the Community*, Teacher's Guide. (Dubuque, IA: Kendall/Hunt Publishing Company, 1988).

Epstein, S. S., L. O. Brown, and C. Pope. *Hazardous Waste in America* (San Francisco, CA: Sierra Club Books, 1982).

Fischer, K. E. "Self Audits of Hazardous Waste Operations in Laboratories," *J. Chem. Ed.* 64 (9) (1987).

Steere, N. V., Ed. *CRC Handbook of Laboratory Safety*, 2nd ed. (Boca Raton, FL: CRC Press, Inc., 1985).

APPENDIX IV

Laboratory Curriculum Revision

McDuffie, T. E., Jr., and J. Anderson. *Chemical Experiments from Daily Life* (Portland, ME: J. Weston Walch, Publisher, 1980).

Shakhashiri, B. Z. *Chemical Demonstrations: A Handbook for Teachers of Chemistry*, Vol. 1 (Madison, WI: The University of Wisconsin Press, 1985).

Summerlin, L. R., and J. L. Ealy, Jr. *Chemical Demonstrations: A Sourcebook for Teachers*, Vol. 1 (Washington, DC: American Chemical Society, 1985).

________. *Chemical Demonstrations: A Sourcebook for Teachers*, Vol.2 (Washington, DC: American Chemical Society, 1987).

Tocci, Salvatore. *Chemistry Around You: Experiments and Projects with Everyday Products* (New York: ARCO Publishing, Inc., 1985).

Wright, Jill D. *Teaching Science Today* (Portland, ME: J. Weston Walch, Publisher, 1982).

APPENDIX V

Government Documents

Connecticut. *Hazardous Waste Management, Regulations* (revised 14 February 1986).

U.S. Environmental Protection Agency, Office of General Counsel. *Clarification of 40 CFR 261.5(g)(3)(iv), Memo* (9 October 1986).

CHAPTER 5

The School Science Laboratory Inventory and Disposal Project in Illinois—Chemical Disposal

Ralph Foster, James P. O'Brien, and Max A. Taylor

THE PROBLEM

The hands-on experience with chemical reagents has always been, and will continue to be, a very important tool in the teaching of the sciences. During recent years, however, studies have shown that certain of these chemicals are toxic and can therefore cause health problems if they are not handled properly. Additionally, there are a number of other chemicals whose inherent characteristics make them a reactive hazard or pose similar dangers. In response to the increased awareness of the toxic and hazardous properties of certain chemicals, properties that make them less desirable for teaching purposes, school curricula changed to utilize other, more acceptable materials. This resulted in an accumulation of unused quantities of less desirable chemicals on school stockroom shelves. Schools often do not have adequate financial resources to properly dispose of these hazardous chemicals. Many sit on the shelf, becoming more dangerous as they age, or lose their labels, making their ultimate disposal more difficult and costly.

RISKS

Many people may not be aware of the hazards of school chemical stockrooms full of old chemicals. Students, school maintenance personnel, and emergency responders are all individuals who could easily be exposed as a result of a chemical storage accident. Children are more susceptible than adults to many chemical toxins. Therefore, proper handling and storage of chemicals is particularly paramount in the school laboratory. Other nontechnical personnel such as custodians, firemen, or ambulance paramedics may be involved in a response to an accident, justifying efforts to minimize the danger to which they might be exposed. By keeping the storage area clear of unnecessary hazards, the potential for dangerous exposure is greatly reduced.

SOLUTIONS

Many of these materials, if put in the trash, could present hazards to trash collection/disposal personnel and introduce toxins into local sanitary landfills that are not designed to adequately contain or detoxify the chemicals. The alternative of flushing them down the drain presents potentially life-threatening hazards to the disposer, as well as to anyone working on the plumbing or sanitary sewer systems, and the toxins introduced could have disastrously adverse effects on the functioning of the local sewage treatment plant. The local environment could be harmed if waters containing the materials pass through the sewage plant untreated. To handle the problem of disposing of excess or no-longer-used chemicals in a responsible and environmentally acceptable manner entails hiring a special contractor to properly dispose of the waste. This is a costly proposition, often running several thousands of dollars, an amount beyond the usual resources of many school budgets.

HOW WE BECAME INVOLVED

The Emergency Response Unit (ERU) of the Illinois Environmental Protection Agency (IEPA) deals daily with spills of hazardous chemicals and the proper cleanup and disposal of the resulting waste and contamination. ERU has been involved with several incidents resulting from old school laboratory chemicals, including a shelving unit collapse, the reaction of two unknown chemicals while still on the shelf, and school fires that resulted in contamination of debris. In addition to these emergencies, we frequently receive letters from teachers requesting disposal guidance and assistance. This increased dramatically between 1983 and 1986. While not a principal function of ERU, we felt that this application of our expertise was worthwhile if it prevented future emergencies. Unfortunately, it was beginning to take larger amounts of time away from our primary role of response and follow-up to chemical emergency incidents. It became apparent that ERU staff alone were not going to be able to adequately address the demand. Moreover, disposal of chemicals that could not be treated in the schools to render them nontoxic or nonhazardous was problematic, as there were no inexpensive options for some materials.

THE PROJECT

Therefore, we explored other ways to address the problem. It was thought that the most efficient method would be a state-wide project encompassing all schools to alleviate a big portion of the problem at one time. By accumulating quantities of chemicals and operating the project on a large-scale basis, much of the administrative, transportation, and other overhead costs could be drastically reduced. The expertise that the chemists of the Emergency Response

Unit had in handling disposal problems and dealing with contractors lent itself well to administering and monitoring a collection and disposal project. This option was also consistent with IEPA's increased emphasis on proactive efforts intended to prevent accidents, rather than on a wholly reactive policy.

QUANTIFYING THE PROBLEM

In order to gauge the scope of the problem in terms of quantities of chemicals and interested schools, a survey was put together and sent out to all of the secondary schools in Illinois (public and private). The results of that survey confirmed that there was indeed a widespread problem, that teachers wanted to dispose of certain chemicals, but lacked the resources to do so properly. Survey results revealed that there were not only hazardous chemicals, but also large quantities of nonhazardous (or easily treatable) chemicals on which teachers were requesting disposal advice. The IEPA chose to separate its approach and categorize all materials to be disposed into two categories. The first one, being hazardous, required special disposal arrangements. The other, being nonhazardous, was easily handled in the schools.

NARROWING THE SCOPE

The survey lists were scanned by ERU chemists to determine which materials should be included in the hazardous waste collection. These hazardous chemicals were then compiled into one list per school indicating materials and quantities to be handled. These lists formed the basis for the collection portion of the project to be handled by a disposal contractor. All of the remaining items were classified as nonhazardous or easily neutralizable. It was decided that to pay a contractor to handle the nonhazardous items would not be cost effective. After investigating an acceptable disposal or treatment method for each nonhazardous material in the literature, a letter explaining appropriate handling and detailing proper disposal or treatment techniques for each was sent to the school. These procedures ranged from treating materials as normal garbage to instructions for performing chemical neutralization reactions. The references we found most helpful in our research are listed at the end of the chapter.

FUNDING

An initial hurdle to be confronted in such an undertaking was how to fund the collection project. Options that were considered included having each school pay their share of the total cost; funding the project from the regular IEPA budget; trying to obtain public or private grants or gifts; obtaining a special appropriation from the legislature; or utilizing state emergency spill cleanup funds. In our case, the use of spill cleanup funds was allowable and less time consuming to arrange than the alternatives. Others attempting similar projects may have to expend a greater portion of their total effort on obtaining funding. After consulting with disposal contractors, and basing our figures on preliminary survey data, we estimated that the job could be accomplished at a cost of one million dollars or less.

TIMETABLE

The amount of time allocated to planning must not be underestimated. For obvious safety reasons, we decided to conduct our collection activities during June, July, and August of 1987 to coincide with summer vacation and to avoid having students around. We had started preparing a Request for Bid (RFB) during the previous January, a five month lead time that was almost too short. We suggest an eight month lead time with five *more* months to conduct a survey, if that is deemed necessary. Of the five months we used, two went for development of the comprehensive 19 page RFB. This took three drafts and several meetings. Another month was given for potential bidders to respond. Two weeks were used to evaluate the bids and select the best one. Three more weeks were then needed to negotiate a final contract and obtain all signatures necessary to make it binding. This left only three weeks for the chosen contractor to prepare in earnest for the collection.

SELECTING A CONTRACTOR

For a project of this magnitude to be done expeditiously, safely, and cost effectively, it was realized that the RFB would have to be detailed, but not so restrictive that an innovative approach would be precluded. The bid document was compiled by representatives of the legal, purchasing, and technical groups within IEPA. Deciding on a Scope of Work and then putting it down in the bid document was the first and foremost concern as bid preparation began. The Scope of Work was set up to include both the pick up of all unwanted materials from the compiled lists (including unknowns) and their proper transport and disposal at an approved disposal facility.

SCOPE OF WORK DECISIONS

Selection of collection sites became an important issue that had to be decided before a detailed Scope of Work could be specified. An option considered was having one collection site for the state. This was regarded as unsatisfactory due to the long distances schools would have to transport materials, discouraging participation and increasing the risk of an accident. The other extreme of having the contractor go to each school for pickup was not considered cost effective, as many schools had less than ten pounds of materials and it would have taken a considerable amount of time to finish the project. We finally decided to have 57 regional collection sites in the pre-existing 57 educational service regions. The contractor was also to be responsible for collecting at those schools that were listed as having over 100 lb of waste chemicals, in order to minimize transport of large amounts by school personnel. Since Illinois had a large number of schools involved in the project (over 700), it was decided to divide the state into four equally sized quadrants. This was done to allow smaller regional contractors to have an opportunity to be awarded part of the contract. Contractors could bid on any number or combination of the quadrants, but could not submit one bid on the state as a whole. In this way, the IEPA was able to evaluate the bids of small and large contractors on an equal basis. It was also thought that dividing the state into quadrants would make it easier to ensure that the work would be completed within the time frame allotted.

EVALUATING BIDS BY CONTRACTORS

The process of analyzing the bids received consisted of evaluating three basic areas—the financial integrity of the contractor, the cost proposal, and the technical quality of the proposal.

The financial integrity question was judged on whether the bidder's cash flow was sufficient to both start and complete our project. This was done by analyzing the bidder's latest annual audited financial statements. The result was a pass or fail rating with failures to be excluded from any further consideration. Additional safeguards were built into the proposal, including a 5% bid deposit, a performance bond equal to the contracted amount, and a 10% withholding on all payments pending successful completion of the project.

The cost proposal included tabulating the various fixed and unit costs requested by the RFB. Because there was some uncertainty regarding the total volumes and types of materials that would be received, unit costs had been solicited. Having costs broken down in this manner reduced uncertainties, but made it difficult to compare bids to each other. Our solution was to create a hypothetical project profile to which the fixed and unit costs were applied. This resulted in a hypothetical total cost for each bid that was then compared to similarly derived totals from other bids. Quite a cost range was observed with the highest bids being twice as much as the lowest bids. Cost was not the only criteria we used to select successful bids, however. Technical quality was also considered.

Specific technical aspects of each bid were rated on a scale of one to ten by each member of an evaluation team. The total averaged technical score was then used to rank the bids in terms of technical quality. The following items were scored: project management, process description, experience of collection station personnel, plans for the arrangement of collection sites, flow chart and narrative of collection station operation, traffic control, instructions to participating schools, vehicle unloading procedures, chemical assessment and segregation, disposal packaging, documentation procedures, unknown characterization procedures, emergency procedures, personnel protection, debris disposal and site decommission, inclement weather contingency, site security, disposal arrangements, descriptions of how different categories of materials would be treated or disposed of (the RFB specified our preference for incineration or other treatment over landfill disposal), the procedures for dealing with unknown (unlabeled) chemicals, direct pick-up procedures at schools with over 100 lb, procedures for handling unexpected materials (not listed on the survey forms), and general site safety and contingency procedures. Of course, the RFB had specifically requested narratives on all those topics and, in many cases, listed minimum expectations.

The goal of this process was to accept a low cost bid with high technical scores. Six firms bid on our project, three of which received comparably high technical scores. Of these, the lowest bid was selected. Not all contractors bid on all quadrants. However, the high score, low bid contractor had bid on all four, and was awarded the whole project contract. This contractor was Chemical Waste Management, Technical Services Division of Alsip, Illinois.

LEGAL ISSUES

As with any complicated contract of this type, it is imperative to have a lawyer review and/or draft it. Specifically, issues to be addressed include liability, termination, methods of payment, contract period, amendments, assignments, subcontracts, responsibility of the contractor, warranty of the contractor, audits and access to records, confidentiality, insurance, indemnity, additional work, funding contingency, and special issues peculiar to the jurisdiction in which the project occurs.

REGULATORY ISSUES

Since specific procedural requirements apply to generators of hazardous waste, a mountain of paperwork would have been required if each school was considered a generator. To minimize the documentation, the state decided to assume generator responsibility by declaring the materials wastes at the collection sites, when they were received and assessed as such by the disposal contractor.

Transportation of hazardous materials is also highly regulated. Regulations require specific packaging, marking, use of specified labels, accompaniment by a specific bill of lading, and placarding of vehicles. This can become very complicated, especially so for school personnel not ordinarily exposed to these procedures. To simplify these procedures, arrangements for an exemption from specific packaging regulations with substitution of equivalent, but more readily available, packaging materials. We prepared explicit instructions on how to package the materials and mark the containers. We enclosed a form to fill out as a bill of lading and included instructions on what to do in an emergency. One of the most confusing parts of this process was assigning materials to specific hazard classes for segregation and marking. We simplified this by researching and assigning all materials reported in the survey to hazard classes. A comprehensive list of these accompanied the packing instructions. Another packaging issue we confronted was what to specify as a sorbant packing material. We decided to insist on granular clay (oil dry or kitty litter) because organic sorbants might react with oxidizers in an accident situation.

LOGISTICS

Selection of an appropriate collection site was an important step with implications for many facets of the total project. The sites were selected taking into account the following criteria:

1. Central location for the total service area (minimize risk due to long distance transport and maximize participation due to small time investment by schools);
2. Waste volume at the collection site (Over 100 lb would save a trip to collect it specially. This would also lessen the risk from transporting large quantities);
3. Size of available facilities (need to have enough parking area to set up and run collection site safely);
4. Accessibility of actual collection area (Try to obtain a site that is easily locatable, but not on a congested thoroughfare or next to the busiest park in town, again for safety reasons).

In all cases, we were able to obtain permission to use a school parking lot as a collection site. Sometimes this necessitated a vote of the local school board, so sufficient lead time should be allotted.

SAFETY AND CONTINGENCY PLANS

After developing the prototype collection site, there were a wide variety of problems that needed to be resolved prior to starting the actual collections. Plans had to be put in place to handle inclement weather, emergencies, traffic, media, site safety, and site security. Although there are many different approaches to solving these problems, we covered them in the following ways. The collections were performed under a tent with a well-defined work area off limits to personnel not wearing protective gear. The contractor was required to develop a general site safety plan, updating it for each collection site with pertinent local information. Each school was assigned a unique time to arrive at the collection site and the traffic pattern was set to allow each vehicle, in turn, to drive up to the unloading area. A queue was formed behind a sign placed a safe distance back from the unloading area. Enough space was allowed for a number of vehicles to be in line without interfering with the normal traffic flow in the area. A well-defined site security and clean up regime was in place to ensure that the collection posed no additional dangers to the neighborhood and that the site was left in a safe and orderly state.

ADVISING AFFECTED OFFICIALS

With a project of this scope and nature, one of the most important tasks is to be certain that all public officials have knowledge of the collection before it occurs. They may express valid concerns that have been overlooked, they may offer local assistance for the project, or they may want to become publicly associated with the positive aspects of the effort. Although the possibility exists that conflicts could arise, early contact will allow time to resolve such problems before the collection day. It is much more difficult to deal with a local problem publicly at the collection site, than privately a few weeks ahead of time. The types of public officials that were contacted in the project fell into the following categories:

- Local Community Representatives
  - Fire Chiefs
  - Political Figures (Mayor, City Manager, etc.)
  - Emergency Coordinators (Civil Defense, Police)
  - School Principals, Local School Boards
- State Representatives
  - Political Figures (Legislators, Senators)
  - State Police (may be responsible for enforcing DOT regulations)
  - Transportation Authorities
  - State School Officials

SPECIAL ARRANGEMENTS

Some materials require special handling beyond what most hazardous waste disposal contractors are used to and are prepared to deal with. Highly reactive or potentially explosive materials are one category, radioactives are another. There are contractors available who specialize in handling one or the other of these types of materials. In our case, we chose the option of handling the problem materials through established expertise and capability within government agencies.

A few schools had low-level radioactive materials that were primarily natural ores, sources for radiation demonstrations, or salts of uranium. These items were identified from the survey, and IEPA staff collected and transported them to a cooperative government laboratory. There they were combined with that laboratory's usual wastestream of low-level research radioactives. Only about 15 lb of such material were found in Illinois schools, making this option acceptable. Interestingly, only one bottle of material collected had a higher activity reading than the calibration source attached to the radiation survey meter utilized (4.0 mr/hr).

Highly reactive materials, along with potential explosives, were destroyed by reaction or induced explosion. These chemicals included sodium, potassium, lithium, calcium, phosphorus, calcium carbide, sodium azide, potassium azide, benzoyl peroxide, sodium peroxide, dry picric acid, old diethyl ether, old isopropyl ether, old dioxane, and old tetrahydrofuran. To minimize the potential for a transportation accident, prior arrangements were made with the Bomb Squad of the Illinois Secretary of State's Police to respond to each individual school as necessary. This squad was experienced in treating reactives according to procedures developed during a long association with IEPA's Emergency Response Unit. Unless closely supervised, explosive disposal can easily become explosive dispersal. We do not recommend the use of bomb squads unless their procedures are carefully scrutinized.

LESSONS LEARNED

Despite the planning and attempts to anticipate and prepare for problems, hindsight allows us to identify some cautions for those planning similar endeavors. We underestimated the administrative effort required, and the schools did not always follow directions exactly. These problems were successfully overcome, but more attention to them during planning appears warranted.

Administration of the project was time consuming and much more involved than was originally thought. Some of the problems in this area were due to the short time frame which had been predetermined at the outset of the project. Determining the final list of materials that would be acceptable in the collection by going through each additional list and then consulting references was very tedious. Inputting these additional chemicals into a computerized database and trying to put together all of the information on a single school, involved long hours of work. All the while, the staff involved was trying to maintain its normal workload. Working with more than one contractor would have significantly increased this problem. In retrospect, we advise that assigned personnel should be involved exclusively in a project of this scale, or that alternatively, more of the administration should be part of the work tasked to a contractor.

Responses on the original survey forms were inaccurate, incomplete, and diverse. They were first tabulated based on 71 anticipated problem chemicals, but after going through all of the additional lists, the number of chemicals included swelled to 680. If this had been originally taken into account, things would have gone more smoothly. Many of the respondents did not understand some of the differences in listed materials, i.e., confusing fuming sulfuric acid with reagent sulfuric acid. Our not specifying what units of measure to use was also an ambiguity that could have been remedied. Sometimes only a value was entered with no reference to what units were used. To add to the confusion, many solutions did not list any concentrations. Many of these problems would have been minimized if time had allowed a confirming inventory to be sent back to the school for approval.

The schools who had been asked to be a collection site and many local officials expressed reservations about being a "hazardous waste dump site." They mistakenly thought that the chemicals were going to be left at the site, rather than collected and removed. Explaining things fully and double checking what the listener understands are very important in a commonly misunderstood and feared subject area.

The letters specifying what is acceptable for bringing to the collection site need to be very explicit. There was a common problem with schools bringing in related materials that were not covered. Items such as pesticides, shop class wastes, and auto maintenance wastes were routinely offered for disposal, at times in drum quantities. These have to be handled in a much different way than the small containers of laboratory chemicals in this project.

Latecomers were a constant problem throughout the project. Their inevitability should be recognized, and plans should be developed to handle the different types of latecomers. Some methods that we used were to direct them to another collection site and to set up another collection effort for those that missed. We scheduled four collection sites in December to accommodate those schools. The very, very late we directed to contractors to make their own arrangements. A more desirable method would be to establish some sort of long term program, such as a once-a-year collection or something similar. Many of the problems that developed in this area were also related to the timing constraints that the project experienced.

The final problem noted was billing and being able to accurately track the costs associated with a particular school. Even though the billings for the project were very detailed, trying to verify the quantity of material brought to the collection site by each school was not practical using the system we had developed. This was because of the large amounts of unanticipated materials turned in at the collections and the logistics of keeping track of those. If we had arranged for schools to turn in a verified and corrected inventory sheet at the collection, this might have been avoided.

ILLINOIS COLLECTION STATISTICS

A total of 27,238 lb of hazardous materials were collected during 85 cumulative days of collection site operations. Seven hundred schools participated. The average school brought in 39 lb of unwanted chemicals, with one school having 633 lb. Forty-six schools had over 100 lb each. These figures do not include those chemicals that were nonhazardous or easily treatable and therefore were not collected for special disposal by the project. Incineration by a hazardous waste incinerator was the disposal method used for 25,373 lb or 93% of the total collected. Recycling was possible for 4%, which was 1,100 lb of elemental mercury. Only 765 lb, consisting of asbestos and mercurial salts, was landfilled as hazardous waste.

The IEPA's administrative effort comprised 1200 hours to plan and monitor this project. The contractor was paid $347,777. Of that total, 48% percent ($167,251) went for administration/collection/mobilization, 34% ($119,103) went for disposal, 11% ($38,608) for unexpected materials, and the testing of 527 unknowns cost $22,815 (7%). The total does not reflect the $7000 cost of the original survey. In terms of unit costs, approximately $13 was paid per pound for collection and disposal (incineration). This averaged about $500 per school.

An important measure of success was that no injuries occurred and no spillages were reported, justifying our efforts to stress safety over convenience and cost.

FUTURE PLANNING

The IEPA *does not* plan to conduct another of these collection projects. We believe that the responsibility to properly plan, purchase, and provide for appropriate disposal of chemical teaching aids is rightfully that of the local schools. This project was intended to allow the schools to start fresh by removing chemicals accumulated before the current awareness of the hazards became prevalent. Accident prevention was our primary motivation for conducting this project. However, the long-term success of that effort depends on the continued commitment of local science teachers and school administrators.

BIBLIOGRAPHY

Bretherick, L. *Handbook of Reactive Chemical Hazards*, 2nd ed. (London: Butterworths, 1984).

Catalog/Handbook of Fine Chemicals, 1988–1989 ed. (Milwaukee, WI: Aldrich Chemical Company, 1988).

Chemical Catalog/Reference Manual, 1985 ed. (Batavia, IL: Flinn Scientific, Inc., 1985).

Gardner, W., E. Cooke, and R. Cooke. *Handbook of Chemical Synonyms and Trade Names*, 8th ed. (Boca Raton, FL: CRC Press, Inc., 1978).

Hawley, G. G. *Condensed Chemical Dictionary*, 10th ed. (New York: Van Nostrand Reinhold Company, 1981).

National Research Council. *Prudent Practices for Disposal of Chemicals from Laboratories* (Washington, DC: National Academy Press, 1983).

CHAPTER 6

Developing a Chemical Management System

James A. Kaufman

INTRODUCTION

The use of hazardous substances in academic science laboratories and the problem of their disposal poses many significant challenges for both educators and their students. The challenges include, but are certainly not limited to, meeting our responsibilities to protect the environment, complying with the law, learning how to handle chemicals more safely, and teaching about hazardous wastes.

The Laboratory Safety Workshop at Curry College has been addressing these challenges for nearly a decade. We are a national center for training and information. We offer a wide range of programs and services. These include seminars on laboratory safety, a 16-page newsletter that is published three times a year by Fisher Scientific-Educational Materials Division, a free audiovisual lending library, and a variety of laboratory safety publications including our *Laboratory Safety Guidelines*.

One of the major themes in our Center's programs and publications is that "Life is Filled with Hazards." Science laboratories are not the only places with hazards. In fact science teachers have a golden opportunity to teach their students about life's hazards and how to protect themselves, and to help them to live safer, healthier, longer lives. Unfortunately, too often they fail to take advantage of this opportunity.

Our Center looks for applications of laboratory hazards and problems in our everyday life. Safe chemical management is one such issue. In one sense, chemicals and hazardous wastes are simply a part of life. It is time we learned to do a better job. The laws are changing and schools need to comply. Sometimes the costs seem high, particularly when one is not used to the idea of paying for chemical disposal.

But it is like the FRAM Oil Filter Company advertisement, "You can pay me now or pay me later." And I am not just talking about money—I am talking about our environment and our health.

Some people believe that companies and communities do not do a good enough job handling hazardous chemicals. Who taught them how to behave! It was our schools, our colleges and our universities.

The problem can be summarized this way: Our lives are filled with chemical hazards. We are not learning how to deal safely with these chemical hazards. We are not preparing our students to work and live safe and healthy lives or how to protect our environment.

The solution to the problem is relatively simple: We need to develop and implement chemical management systems. We have an opportunity that we often as teachers, educators, and concerned citizens fail to take advantage of—to teach children how to identify hazards and how to protect themselves and the environment.

Because life is full of hazards, we need to *make safe chemical management an integral and important part of science education, our work, and our lives.*

THE CHEMICAL MANAGEMENT SYSTEM

A chemical management system is a complete program to deal with hazardous substances. The system is composed of nine elements:

1. Assuming Responsibility
2. Determining Hazards
3. Inventorying Chemicals
4. Purchasing Philosophy and Policy
5. Handling and Storage
6. Avoiding Waste Formation
7. Disposing of Wastes
8. Obeying the Law
9. Teaching Home Application

The discussion of these nine elements form the core of a one-day seminar that our Center presents on this topic. In this chapter, I would like to comment briefly on each of these, with special emphasis on assuming responsibility and teaching home application.

ASSUMING RESPONSIBILITY

As far as I am concerned, this is the main issue. Administrators, school boards, and trustees at schools and colleges need to decide that chemical management is important and that someone will be responsible for seeing that the problem is addressed.

If nothing else happens, I hope to convince readers that the administration of academic institutions must establish policies for the safe disposal of chemicals. They must provide adequate funding to implement these policies.

A sample institutional hazardous materials policy might look like this:

> It is the responsibility of the institution and its staff to ensure that our educational programs and other activities use, store, and dispose of hazardous substances in ways that protect and promote the health and safety of our students, our staff, and the environment.

These issues do not get attended to on their own. Someone must be appointed either coordinator of the school's chemical management system or hazardous waste coordinator. Someone needs to have the time and the resources to address the problems. With some institutions facing a declining enrollment, this may be a very good time to put a part of someone's teaching load into coordinating this area. It definitely should not be just one more uncompensated responsibility.

Generally, schools must operate as one site; science laboratories, art studios, vocational shops, and building and grounds all produce hazardous wastes. The U.S. Environmental Protection Agency (EPA) says that schools should be treated as a single unit. The exception to this occurs when the facilities are divided by public streets and roadways.

On the other hand, many school districts have gotten permission to treat the whole district as a single generator. They transport their wastes within the school system to a single location and dispose of them from there.

Having a coordinator is not going to be sufficient. Everyone needs to be responsible: school boards, superintendents, principals, department heads, teachers, students, authors and publishers. Having a coordinator does not mean that there is less individual responsibility. Everyone generating wastes needs to work with the coordinator to have a safe workplace and protect the environment.

DETERMINING HAZARD

Once a coordinator has been appointed and policies have been established, it is necessary to determine which substances are actually considered hazardous. For this it is necessary to be familiar with the lists and definitions of hazardous materials in state and federal regulations. An excellent summary of these is contained in Chapter 1.

In general substances are considered hazardous if they are specifically listed by name or have the characteristics of a hazardous material, i.e., they are corrosive, ignitable, reactive, or toxic.

These listed and defined materials may not be the same as what might be expected by either general health and safety considerations or the definitions and regulations of state and federal right-to-know laws. It is necessary to refer specifically to hazardous waste regulations.

INVENTORYING CHEMICALS

The next step is generating a complete inventory of all the chemicals used or stored at an institution. The coordinator needs to enlist the assistance of all departments to review their own areas and prepare a list of the names, amounts, and location of their chemicals.

It is helpful to provide them with a summary of the kinds of substances that are liable to be hazardous. This will help to avoid missing some of the things that sometimes get overlooked because they are taken for granted.

Consider implementing a computer based inventory system. This will simplify producing lists of chemicals for distribution to each department. It will also facilitate developing chemical exchange programs. Several excellent computer programs are available from Fisher Scientific-EMD and Flinn Scientific, Inc.

PURCHASING POLICY AND PHILOSOPHY

It used to be taken for granted that buying the larger size container would be more economical in the long run. This is no longer true. By the time the cost of disposal is included, it is better to purchase what can be used in a reasonable period of time.

The American Chemical Society publication "Less Is Better" makes this point most clearly. Today, nearly 40% of disposed laboratory chemicals are unopened, unused bottles. Buy only what is needed and do not accept unusable "gifts."

Get in the habit of checking the institution's master inventory to see if someone else has what is needed. Perhaps they would be happy to give it away.

Establish a written purchasing policy that summarizes these concepts and emphasizes buying with the idea of eliminating disposal costs.

HANDLING AND STORING CHEMICALS

The safe handling of chemicals begins with being able to answer four simple questions:

1. What are the hazards?
2. What are the worst things that could happen?
3. What do I need to do to be prepared?
4. What are the prudent practices, protective facilities, protective equipment, and emergency equipment needed to minimize the risks?

If these four questions cannot be answered, chemicals should not be handled. A good general introduction to this subject is contained in the American

Chemical Society's *Safety in Academic Chemistry Laboratories*. Single copies are free (1155 16th Street, N.W., Washington, DC 20036).

Proper storage of chemicals means that the storage area is kept locked, and unauthorized persons do not have access. In addition, there is sufficient space for the containers. They are no more than two deep on the shelves and there is room between containers to reach in and remove the one wanted.

One inexpensive way to gain extra space is to carefully combine (in a fume hood) containers of the same material that are half-filled or less. Check to be sure that the two materials are, in fact, the same. Do not work alone.

The storage room should be properly ventilated. The rule of thumb is one cubic foot of air per minute per square foot of floor space. And the ventilation should be 24 hours a day, every day. A recommended minimum rate is 150 cubic feet per minute.

The storage room should have appropriate fire protection. This may include fire extinguishers or suppression systems, explosion proof lighting, grounding facilities, and heat and/or smoke alarms.

The National Fire Protection Association (Batterymarch Park, Quincy, MA) publishes *Fire Protection for Laboratories Using Chemicals – Code #45*. This is a most useful guide for understanding and meeting fire protection needs.

Last on my list of storage criteria is arrangement. Certain basics need to be observed. There are five major storage categories: flammables, corrosives, toxics, reactives, and non-hazardous materials. They should be kept in separate sections of the storage area. Flammables need to be kept in approved storage cabinets. Fuels should be separate from oxidizers. Acids and bases should not be together and nitric acid should be by itself. Water reactive materials need a cool, dry place.

Two free posters on chemical storage are available from Fisher Scientific-EMD (4901 W. LeMoyne Street, Chicago, IL 60651).

AVOIDING WASTE FORMATION

There are seven general methods for avoiding waste formation.

1. Surplus chemical exchange has already been mentioned in several places. It certainly is worth repeating. What one user does not need may be exactly what someone else in the same institution or the one down the road badly needs. Make a list of what is not wanted and circulate it to other departments and institutions.
   Chapter 12 describes some of the more well-established waste exchange programs and their impact on disposal costs.
2. Recycling can be applied in two ways. Spent solvents can be recovered by careful distillation, and curricula can be developed so that the products of one experiment become the starting materials of the next. The copper cycle experiment is one good example.

3. Substitution involves replacing more hazardous materials with less hazardous ones. Can toluene be used in place of benzene?
4. Microscale reactions are becoming increasingly popular in both school and college laboratories. Equipment costs are lower, breakage is lower, chemical costs are lower, and there is less waste and pollution.
5. Volume reduction means not putting half-filled bottles in lab pack containers. If we pay per drum, we are paying to dispose of air! It also means not disposing of dilute aqueous solution when it is possible to precipitate the toxic ions and thus have only the solids to dispose.
6. Dilution is using a 0.01 M solution instead of one that is 1.0 M. Another aspect of dilution is checking with your local water treatment facility to determine exactly what materials they can and will treat effectively.
7. In-house conversions involve having the time, expertise, and facilities to properly treat materials to make them less hazardous. Section V of this book is devoted to this subject. It is possible to do this very effectively and economically.

DISPOSING OF WASTES

There are several points to keep in mind concerning the actual disposal of wastes.

Each hazardous waste generator needs to obtain an EPA Waste Generator Number from the EPA or their state agency.

Once a central coordinator has been designated, the coordinator needs to:

1. Identify all sources and types of hazardous wastes at an institution.
2. Calculate the total volume of hazardous waste generated monthly. This makes it possible to determine in which category of waste disposer an institution belongs.
3. Segregate wastes at the point of generation.
4. Establish routine storage, disposal and record keeping procedures as well as emergency plans.
5. Establish a waste minimization program.

The selection of a disposal company is an important decision. It should be done with considerable care. A waste generator may be liable for the company's mistakes. This is what is meant by "cradle-to-grave responsibility."

Check their reference and permits. Find out where the waste is going. Consider forming a consortium to reduce the costs. And lastly, talk with the disposer about things that can be done to help reduce the expense.

OBEYING THE LAW

It is an institution's responsibility to ensure that state and federal laws are obeyed. The hazardous waste coordinator should obtain copies of the laws and become familiar with them. Special training programs and reference materials will no doubt be helpful.

Four books that I have found to be particularly helpful are:

1. *Handbook of Hazardous Waste Management for Small Quantity Generators*, R. W. Phifer and W. R. McTigue, Jr., Lewis Publishers, Inc., 121 South Main Street, Chelsea, MI 48118, 1988.
2. *Hazardous Waste Management at Educational Institutions*, National Association of College and University Business Officers, One Dupont Circle, Washington, DC 20036, 1987.
3. *Hazardous Waste Management Handbook for Schools*, Clean Harbors of Natick, 15 Mercer Street, Natick, MA 01760, 1988.
4. *Understanding the Small Quantity Generator Hazardous Waste Rules*, U.S. EPA, Office of Solid Waste and Emergency Response, Washington, DC, 20460, 1986.

TEACHING HOME APPLICATION

At the same time that schools are developing programs to address the problem of hazardous wastes in their institutions, they should be incorporating these issues into their curriculum. After all, who is going to be dealing with these problems tomorrow?

What will our children find most valuable in our science classes . . . Mendel's genetics, Kepler's laws of planetary motion, Mendeleev's periodic table. I doubt it. We may have learned about these three gentlemen when we were in school, but unless actively engaged in the sciences, it is unlikely that any of this is important to us today.

As educators we are missing out on a tremendous opportunity to teach children about hazards, how to protect themselves, and how to live safer, healthier, longer lives.

Hazardous waste management is just one of many examples of a school chemical issue that relates directly to the home and the rest of student's lives. The average family of four produces about 7,500 lb of hazardous wastes annually. Hazardous waste management should be made part of the curriculum.

Hazardous Wastes From Homes is a good curriculum guide produced by the Enterprise for Education (1320A Santa Monica Mall, Santa Monica, CA 90401). It is a wonderful, full-color 8-1/2x11 booklet that presents a history of the problem, explains about hazardous substances, describes the new methods of treatment, and explains what citizens can do at home and in their communities.

Another very useful item is the *Household Hazardous Waste Wheel*. It is distributed by the Environmental Hazards Management Institute (P.O. Box 283, Portsmouth, NH 03801). Dial the household hazard (paints, automotive, pesticides, etc.) and find the hazardous ingredients, precautions, proper disposal method, and a less hazardous alternative.

CONCLUDING REMARKS

I would like to conclude with two thoughts. The first brings us back to the importance of education. As educators, we need to take advantage of a tremendous opportunity to teach children about hazards, how to protect themselves, and how to live safer, healthier, longer lives.

The last thought is simply this:

> *No task or lesson is so important and no job so urgent that we cannot take time to manage our chemicals safely.*

I hope we will all take the time to make safe chemical management an integral and important part of science education at our institutions and in our lives.

SECTION III

Identification of Unknown Chemicals

CHAPTER 7

Characterization of Unknown Laboratory Chemicals for Disposal

Stephen R. Larson

INTRODUCTION

What is an unknown laboratory chemical and how are these materials generated? Every place in an institution where chemicals are used or processed is a potential point of generation for chemical wastes. There are many terms for "unknowns" including uncharacterized chemicals and unlabeled chemicals. In *Prudent Practices for Disposal of Chemicals from Laboratories* (National Academy Press, 1983), unknown laboratory chemicals are referred to as orphan reaction mixtures.

Packages, containers, and cylinders of solid, liquid, or gaseous chemical substances of unknown composition are present anywhere chemicals are used for several reasons:

1. Absence of organizational policy requiring standardized labeling of all containers of chemical substances.
2. Persons leaving the organization are not required to identify all unlabeled containers prior to formal separation.
3. Labels contain inadequate information or contain unique symbols understood only by the user.
4. Labels have deteriorated or are missing completely due to spillage, leaking, or aging. Spillage can cause obscuring of print on labels as well as corrosion of the label, typically composed of paper. Leakage from the container causes similar effects as spillage. Aging dries glues, adhesives, and paper and causes labels to fall off completely.
5. Substances are transferred to unlabeled packages, containers, or cylinders, and the original containers are untraceable.

POLICIES FOR REDUCING THE FREQUENCY OF UNKNOWN CHEMICALS

Clearly, the most cost effective manner to deal with the problem of unknown chemical substances is to reduce or prevent their generation in the first place. By understanding how unknowns are generated, it is possible to reduce their frequency by implementing policies in the organization and then conducting routine audits to evaluate compliance with these policies.

Policy 1. As a minimum, all labels must list the formal, chemical names of one or more substances in a container.

Policy 2. A standard chemical substance label, self-sticking and plastic coated for chemical resistance should be provided to all users.

Policy 3. All containers used for permanent storage of chemicals should be labeled with the standard chemical substance label. (Some containers are temporary and generally used only in the presence of the user.)

Policy 4. All persons who use chemical substances must have their work area approved by the Hazardous Waste Coordinator prior to separation from the organization.

Policy 5. A list of time sensitive and temperature sensitive chemicals should be distributed to all users. Dates of acquisition and recommended date of disposal should be written on the label. Ethers and other peroxide-forming chemicals are considered time sensitive chemicals, since slow reactions with air over the liquid cause peroxide formation.

A UNIVERSITY HAZARDOUS WASTE PROGRAM

These policies arose from experiences gained while managing a hazardous waste program at a large, private university located in an urban area of a major northeastern city. A formal hazardous waste program began in early 1980. During the seven year period ending in 1987, approximately 18,000 units of hazardous chemical waste were collected and prepared for shipment to approved hazardous waste treatment and disposal facilities. During the first five years of the program, units of unknown chemicals were generated at a rate of 1 per 100 units of waste. A unit of chemical waste is defined as each separate and distinct package or container. These containers range in size from:

micro	0-500 mL
standard	1-4 L
pilot plant	5-55-gal drum
plant scale	more than 55-gal drum quantities

During the last two years of the program, the rate of generation of unknowns was reduced to 1 per 500 units of waste. Nevertheless, by 1987, there were 150 units of waste of "unknown character," potentially hazardous in storage, in the accumulation facility. These units accounted for as much as 25% of the storage capacity of the accumulation facility.

Problems and Solutions

In 1987, a decision was made to eliminate this collection of unknowns. There were three possible methods to solve the following problems:

Problem 1. How does one safely open aged containers of chemicals, some potentially reactive or explosive?
Problem 2. How does one draw a representative sample from chemicals that may be heterogeneous in composition, that is, composed of two or more liquid phases?
Problem 3. How completely must each sample be characterized?
Problem 4. How does one instruct an analytical laboratory to conduct a series of tests consistent with protocols established by treatment and disposal firms?
Solution 1. Establish an in-house dedicated laboratory to analyze unknowns.
Solution 2. Develop an in-house borrowed capability working with faculty and graduate students in the Department of Chemistry.
Solution 3. Use an outside contractor who will:
1. Sample on-site and remove all samples for detailed lab testing.
2. Sample on-site and analyze on-site.

Solution 1 was eliminated because of high, up front equipment and supplies costs. Solution 2 was eliminated because of the liability in opening potentially reactive or explosive chemicals. Solution 3 was accepted and the contractor was required to open, sample, and analyze the materials on site.

Visual Inspection of the Container

It was established that much information about each unit could be assessed through observation of the container. The following information may be gleaned through examination:

1. Examine the physical appearance of the container:
   size of the container;
   materials of construction;
   evidence of container degradation;
   chemicals leaking from the inside, particularly around openings.
2. If the container is transparent, chemical contents can be assessed as to:
   color;
   homogeneity, is there phase separation?
   weight;
   volume;
   presence of crystalline material;
   presence of sediment.
3. If the container is opaque:
   Is the container unique to particular chemical substances?
   Is the material in the container likely to be original to the container?

4. Is the container bulging, or does it exhibit any other signs of internal pressure buildup from gases?

Of 150 units of unknown chemicals, two-thirds were liquid and one-third was solid, as assessed by visual observation and careful "hefting" of the container. The normal waste that is identified waste occurs in consistent ratios month after month. In all likelihood, the unknowns should follow a similar pattern.

- 32% flammable/combustible/ignitable
- 24% corrosive
- 12% oxidizer
- 13% other

Upon later analysis, the ratios of unknowns by type was found to be consistent with the identified ratios.

On-Site Screening Tests

According to the definition of "hazardous waste" as defined by the U.S. Environmental Protection Agency (EPA) in the Resource Conservation and Recovery Act (RCRA), a waste is hazardous if (1) it is listed as a hazardous waste and (2) it exhibits one of the following properties:

1. ignitable (flash point)
2. corrosive (pH)
3. reactive (air or water sensitive or cyanide or sulfide containing)
4. extraction procedure (EP) toxic

Based on the information gathered by the generator, the contractor segregated the 150 units into two groups:

- low risk—likely to be unreactive
- high risk—likely to be reactive or explosive

Wearing personal protective equipment and conducting the operation in a fume hood, the contractor opened each low risk container and removed a representative sample or samples, depending on the presence of two or more solid or liquid phases. Each sample was tested on-site for flammability, pH, and the presence of cyanides and sulfides. Ignitable chemical units that were found to be homogeneous liquids were consolidated into one container for additional "fingerprinting."

Using special apparatus, high risk containers were opened and sampled at a distance. Free liquids were tested for the presence of peroxides.

A second sample was removed from each sample or consolidated samples which were taken to the laboratory and analyzed for inorganics, metals such as

As,Pb,Hg,Cr,etc. and organics, a group of persistent pesticides and polychlorinated biphenyls.

During the operation, the actual opening, sampling, and analysis (screen) time was 1/2 man-hour per sample. By comparison, identified samples are segregated and packed into laboratory waste drums at a rate of 5 minutes per unit.

Several environmental analytical laboratories were contacted to provide quotes on the "RCRA Analysis" of 150 units of chemical waste. The quoted prices ranged from $500/sample to $1000/sample. Most analytical laboratories expressed resistance to opening containers and preferred to leave that step to the generator. It was therefore decided to use the analytical chemistry services of a hazardous waste treatment and disposal firm. The final cost per sample for analysis and disposal was $150/unit.

CONCLUSION

In summary, it is possible to significantly reduce the occurrence of unknowns through the establishment of policies and audit procedures. However, when "unknowns" occur, it is possible to establish a protocol to pre-evaluate each unit through careful visual inspection and "hefting" and to conduct satisfactory screening tests on-site to avoid problems with transporting potentially reactive or explosive chemicals either to an analytical laboratory or to the disposal/treatment site.

CHAPTER 8

Waste Disposal in an Academic Laboratory: Headaches and Solutions

Iclal S. Hartman

THE PROBLEM

About two and a half years ago, the problem of storage of old, unused chemicals and the need for their disposal had reached a state that required immediate attention in our Massachusetts college chemistry department. Along with other academic laboratories, we had been casual, if not careless, over the years. We ordered large quantities when small would do, reordered without checking the stock, opened new bottles before others were used up, and, in general, stockpiled bottles of chemicals, both knowns and unknowns. But with strong state and federal regulations becoming or about to become laws, Occupational Health and Safety Act (OSH Act), right-to-know, etc., we realized that we had to streamline our department, stockroom, and teaching and research laboratories.

We began to look into the necessary steps for disposal of excess and waste chemicals, and started to talk with the representatives of various chemical and hazardous waste transport and disposal companies. We became aware, first, of the magnitude of the cost of such disposal, and secondly, of the amount of work that we still had to do sorting, cataloging, labeling, and packing the chemicals before they could be picked up by a private waste disposal agency. For example, we were asked to provide a description for every unknown or uncertain compound as shown in Figure 1.

We also had to supply the company with the total amount and number of containers of each substance. All this, and we still had to pay thousands of dollars for the chemicals to be carted away, since Massachusetts allows almost no chemical disposal within the state. Hence, we would have to pay the extra expense of out-of-state transportation and disposal, to say nothing of the cost of identification of the unknowns.

ORGANIC OR INORGANIC
ACIDIC OR BASIC - how much acidic, less than pH 2, or above pH11
REACTIVE WITH WATER
FLAMMABLE
FLASH POINT - one to five seconds
EXPLOSIVE
OXIDIZER
HEAVY METAL CONTENT - identity of the heavy metal, if possible
POISONOUS

Figure 1. Information requested for every unknown.

THE SOLUTION

It appeared to us that if all of this work had to be done, perhaps we, as a department, could go the next step and attempt to analyze and identify as many of the unknowns as possible. This would enable us to use our collective wisdom and expertise without creating an undue burden on any one person. We certainly did not want to add this to the faculty's regular work load.

Ours is a small, congenial group of chemistry faculty, and most of us are willing to do extra work, and be flexible about what we should and should not do. Even so, some had objections to this project on the grounds that it was not proper for us as chemistry faculty to undertake such a task. However, although it was to be a voluntary undertaking, we needed as much variety of experience as possible, or we could not hope to accomplish much. So we knew that we either had to do it collectively or not at all. After some discussion, we reached a compromise, deciding to limit our efforts to a maximum of three days. We would analyze as many chemicals as we could during that period. We chose a day at the end of May after final examinations, but before we had scattered for vacation or individual research projects.

ORGANIZING

Collecting and Classifying Unknowns

Before this date, however, a fair amount of organization was necessary. Several weeks before the May date, the director of our lower level laboratories, together with the stock room attendant and paid student help, sorted out all of the knowns we did not need, as well as the unknowns and the uncertains. We had an idea of the contents of the uncertains, but not enough to indicate on the label. The staff marked the location where each unknown and uncertain came from, the organic or inorganic section of the stockroom or a particular teaching laboratory, etc. Any portion of the remnants of a label was saved. If there was a homemade label, for example, it might help us identify the contents by showing what course or research project a substance had been used for. Even if a piece of a remnant label contained only the first letter of the name, A for example, it could help us identify the contents as Arsenic versus Aluminum or Ammonium. All the unknowns and uncertains were assembled in an isolated corner of the stock room. From time to time, faculty members would examine the collection of bottles, attempting to identify the contents visually. It is amazing how many small clues helped us in this part of the work. Such things as the shape or color of the bottle, the writing on the label, and the date or course number were all clues that often led to positive identification. One interesting case invovled the contents of a large bottle, irregularly shaped and sized lavender crystals. A preliminary guess as to the identity of the contents was litmus. However, a simple acid test ruled that out. Upon closer examination of the lavender crystals, two of us recalled a freshman crystal growth experiment and identified the crystals of a chromium-aluminum salt.

Those bottles identified were then either removed to the known pile to be discarded or more often placed back on the stockroom shelves. Word was sent to all students and faculty to bring any unused, unknown, excess (hoarded or unhoarded) chemicals to be similarly processed and added to our inventory. We went through the upper level teaching and research laboratories, collecting any bottles unlabeled with the name of either a student or faculty member. On the side we helped the biology department with some of the chemicals they were discarding by helping them make a correct list. Their list had abbreviated the names of the compounds, which cut off the endings. Thus, a "sulf" could have been a sulfide, sulfite or sulfate!

After all of this was done, we isolated any hazardous looking unknowns for external analysis. We simply did not want to handle any dangerous chemicals. We found over five pounds of unused sodium metal, ordered year after year. It probably had a protective oxide layer, but we packed it in shock proof containers and securely taped them away.

Pre-Analysis Preparation

Then, based on our consultation with all the chemical waste companies, we made analysis forms and gave each unknown a number (Figures 2 and 3).

Before the day of the group analysis, we prepared some standard solutions—no fancy, unique reagents, just the usual acids and bases and simple qualitative analysis solutions. We also assembled pH paper, platinum wire for flame analysis, etc. In addition, in the preceding weeks, we did a quick screening of the unknowns into organic and inorganic using a Kofler hot stage, a metal plate with a shallow well for melting point determinations of microscopic samples. For this, the temperature was set to 200°; those melting below were considered organic and those above inorganic.

The Analysis Procedure

When the day arrived for the group analysis, the unknowns were brought to the large freshman laboratory. The student lockers were opened to give us

UNKNOWN #:
AMOUNT:

DESCRIPTION:
FLAMMABILITY & FLASH POINT:
BURN? RESIDUE?
MELT? RESIDUE PLUS WATER?
DECOMPOSE? DISSOLVE? pH?
MELTING POINT:
FLAME TEST:
SOLUBILITY:

	SOLUBLE	INSOLUBLE	COLOR	DENSITY	pH
WATER					
ETHANOL					
DIL. NaOH					
DIL. HCl					
Conc. H_2SO_4					

OTHER TESTS
TENTATIVE CONCLUSIONS:

Figure 2. Form for preliminary analysis of unknowns.

access to glassware. Almost naturally we separated ourselves into four groups:

1. One group did the background analysis (Figure 4).
2. The stockroom attendant did all the melting point determinations.
3. Organic chemists attempted the identification of the tentative organic unknowns. In general, we measured solubility in water and ether and Acid/Base behavior with dilute HCl, NaOH, $NaHCO_3$. We used the Aldrich catalogs (of IR and NMR spectra) as a reference for commercial organic compounds. We looked at IR spectra for C=O, OH, and NH, and at NMR spectra for complexity of CH patterns. For example:
   a) an unknown was:
      - Neutral, insoluble in H_2O;
      - IR shows C=O and N-H typical of primary amides;

UNKNOWN #:
AMOUNT:

I.R. SCAN:
NMR:
OXIDIZABILITY: DILUTE PERMANGANATE
STARCH-IODIDE PAPER
HALIDES: ALCOHOLIC SILVER NITRATE
CYANIDE:
HEAVY METALS:
FURTHER TESTS:
CONCLUSIONS

Figure 3. Form for confirmatory analysis of unknowns.

APPEARANCE
COLOR
FLASH POINT (1-5 SEC)
FLAME TEST
SOLUBILITY
HEAT OF REACTION (WITH WATER)
pH
ORGANIC OR INORGANIC (TENTATIVE)
OTHER CONCLUSIONS (TENTATIVE)

Figure 4. Background analysis.

- NMR shows only two kinds of protons that have low field chemical shifts. CH_2 is typical of Br-CH_2-C(=O);
- Melting point agrees with $BrCH_2CONH_2$;

The unknown is identified as α - Bromoacetamide.

b) Quart or gallon bottle indicates the unknown is a solvent-type of compound:
IR shows OH;
NMR shows two kinds of H and no OCH_3 or OCH_2CH_3;
Water soluble;
High boiling.

The unknown is identified as diethylene glycol.

4. By skipping around the qualitative scheme and zeroing in on the suspected cations (Figure 5), inorganic and physical chemists analyzed the tentative inorganic unknowns, mainly for heavy metals and cyanide.

Our three-day experience was pleasant, bringing back memories of graduate school. Some of our students came by, and intrigued and interested by what they saw, offered to help. We declined for mainly legal reasons.

We had one "near miss" early in the morning of the first day. The contents of one bottle were flat, flaky, cream colored crystals. The comment on the preliminary analysis sheet was that it looked like a student preparation. It also showed that the substance had dissolved in water giving a pH of approximately 2-3. In addition, the preliminary form indicated that when the unknown was added to water, the solution became warm to touch. The physical inorganic chemist receiving it did a few tests and decided it was safe to dump it down the

FLAME TEST:
SOLUBILITY IN WATER:
IF SOLUBLE—CHECK FOR pH
  CHECK FOR CL^- ppt (Ag^+, Hg^{2+}, Pb^{2+})
  CHECK FOR OH^- ppt (above, Fe^{3+}, Cr^3,Al^{3+})
  CHECK FOR SO_4^{2-}ppt (Ba^{2+},Ca^{2+})
IF INSOLUBLE—TRY TO DISSOLVE IN ACID OR BASE
TAKE OBSERVATIONS FROM ABOVE AND APPLY LOTS OF INTUITION TO SUGGEST OTHER SPECIFIC REACTIONS TO TRY.

Figure 5. Cation identification scheme.

sink. The next thing we heard was a big bang and acrid fumes filled the laboratory, which we had to leave for a few minutes until the air cleared. Fortunately, only a small portion of the contents had gone down the sink, for the unknown turned out to be P_2O_5. We concluded that the top of the contents had been in contact with moisture, hence the relatively low heat of reaction upon dissolution in water. We made a note to sample all subsequent unknowns from the middle and the bottom of the containers, and not be too hasty to throw them down the sink.

The process of quickly identifying an unknown without actually going through the individual steps really needed the experienced eye of a good chemist. What younger chemists who think that sodium chloride is a greenish yellow gas might do, we do not know. We had to see, feel, and smell our way through, and enjoyed this immensely. We had one physical chemist, a crystallographer, who upon examining the crystals of the unknowns, could make accurate predictions and quickly confirm them. For example, he knew KIO_4 from its tetragonal clusters and sodium sulfite (Na_2SO_3 from its fine granular, but free flowing properties, easily confirmed by acidification and the subsequent release of SO_2. Cr(III) Nitrate he identified from its black-purple, coal-like appearance, and the fact that when dissolved in water, it showed a characteristic red/blue dichroisim. Its identity was confirmed by the addition of NaOH and the formation of a precipitate, followed by the dissolution of the precipitate in excess NaOH, giving a bright green solution of $Cr(III)\ (OH)_4^-$. In another instance, an unknown had nicely shaped clear, white prismatic crystals. He observed that the crystals lost water readily, resulting in a coating of white powder. They did not effloresce and fall apart the way other hydrated crystals might do. These facts indicated to him a specific hydrate, sodium thiosulfate pentahydrate. He added acid, and the smell and turbid suspension of S confirmed the prediction. It may have been a potassium or a sodium salt of this thiosulfate, but for our purposes it did not matter what cation it was. Our senior organic chemist could zero in on the functional group and the general identity of a compound, and by comparing it with a known or giving it a quick IR scan, could nail it down. We all had a good idea of what chemicals were used in courses and what in research, so the analysis process was not as difficult as when one works with complete unknowns. Collective experience, besides being fun, was necessary for on the spot consultations, for example, getting a fresh eye to look at an unknown.

CONCLUSIONS

In less than three days of work, we reduced the number of unknowns from a few hundred to a few dozen. Those identified either went back on the shelf, mainly for student experiment use, were poured slowly down the sink with lots of cold water, or were placed aside for pick up by the waste disposal agency. We attempted no controlled combustion of any organic compounds. In getting the disposal companies to take away the chemicals, even partial identification was most helpful. Some of the complete unknowns are still waiting to be picked up, and the cost for such disposal is going up.

The administration was grateful, however briefly, for money saved. Now we are doing sensible things such as ordering small amounts of substances in small individual bottles, keeping track of labels (especially in the summer when the stockroom gets hot and labels fall off), recycling solvents, using smaller amounts of chemicals in student experiments, and, in general, staying on top of waste management in our academic laboratories.

SECTION IV

Methods for Treating and Handling Wastes

CHAPTER 9

Tested Laboratory Disposal Methods For Small Quantities of Hazardous Chemicals

Margaret-Ann Armour

INTRODUCTION

Our increasing awareness of the potential damage to human health and to the environment of inappropriate discarding of hazardous chemicals has led to stringent regulations. As a result, the disposal of small quantities of a wide variety of these chemicals presents an expensive and difficult problem. An alternate solution to the problem is conversion of the chemical by a laboratory reaction to nontoxic and nonhazardous residues. Ideally, these procedures should be performed at the laboratory bench by the worker who has been handling the waste or surplus chemicals. In some institutions, the wastes are collected and a specially trained employee carries out the reactions. Over the past six years, a research group at the University of Alberta has been developing and testing in the laboratory methods for the safe and practical conversion of small quantities of hazardous chemicals to nontoxic and nonhazardous residues. The criteria for a suitable reaction are that at least 99.9% of the starting material must be destroyed, and nontoxic and nonhazardous products formed. For most reactions, the products are identified; where this is not practical, the products are tested for mutagenicity using the Ames test.

Appropriate procedures are documented in detail with precise quantities and conditions. A selection of the developed procedures is presented in this chapter.

It should be noted that the disposal of hazardous chemicals, by whatever method, must be performed in accord with local, state, provincial, and federal regulations. Furthermore, these procedures should only be performed using appropriate personal protective equipment and protective facilities. For more information, the reader is referred to *Prudent Practices for Handling Hazardous Chemicals in Laboratories*, National Academy Press, 1981.

HEAVY METAL SALTS

In many states and provinces, the disposal into landfills of solutions containing heavy metal salts is banned. The metals can be precipitated as insoluble salts that are acceptable for disposal. The insoluble salt of choice has often been the sulfide. This requires the use of hazardous reagents such as hydrogen sulfide, sodium sulfide, ammonium sulfide, or thioacetamide. Instead of the sulfide, lead, cadmium and antimony ions can be precipitated as silicates. These salts show similar solubility properties to the sulfides in neutral, acidic and basic aqueous solutions.

Lead Salts

$$Pb^{2+} + Na_2SiO_3 \longrightarrow \underset{\substack{\text{lead silicate} \\ \text{(insoluble)}}}{PbSiO_3} + 2Na^+$$

The soluble lead salt (0.04 mole, e.g., 11 g of lead chloride) is dissolved in water (200 mL) and a solution of sodium metasilicate ($Na_2SiO_3 \cdot 9H_2O$, 34 g, 0.12 mole) in water (260 mL) is added with stirring. The precipitate is collected by filtration, or the supernatant liquid can be allowed to evaporate in a large evaporating basin in the fume hood. The solid is allowed to dry, and is then packaged and labeled for disposal in a secure landfill.

For dilute solutions of lead salts, the sodium metasilicate solution should be added until there is no further precipitation, and the solution allowed to stand overnight before collecting the solid by filtration or allowing the liquid to evaporate.

Cadmium Salts

$$Cd^{2+} + Na_2SiO_3 \longrightarrow \underset{\substack{\text{cadmium} \\ \text{silicate} \\ \text{(insoluble)}}}{CdSiO_3} + 2Na^+$$

The soluble cadmium salt (0.05 mole, e.g., 9 g of cadmium chloride) is dissolved in water (200 mL) and a solution of sodium metasilicate ($Na_2SiO_3 \cdot 9H_2O$, 25 g, 0.087 moles) in water (200 mL) added. A white precipitate of cadmium silicate forms immediately. The mixture is heated at 80°C for 15 minutes to complete the reaction, is cooled, and the precipitate collected by filtration, or the supernatant liquid allowed to evaporate in a large evaporating basin in the fume hood. The solid is dried, packaged, and labeled for disposal in a secure landfill.

For dilute solutions of cadmium salts, the sodium metasilicate solution should be added until there is no further precipitation, and the solution allowed to stand overnight before collecting the solid by filtration or allowing the filtrate to evaporate.

Antimony Salts

$$2Sb^{3+} + 3Na_2SiO_3 + \longrightarrow \underset{\text{(insoluble)}}{\underset{\text{antimony silicate}}{Sb_2(SiO_3)_3}} + 6Na^+$$

The soluble antimony salt (0.02 moles, e.g., 4.5 g of antimony trichloride) is dissolved in water (150 mL) and the solution heated to 80°C. To this hot liquid is added, with stirring, a solution of sodium silicate ($Na_2SiO_3{\cdot}9H_2O$, 15 g, 0.052 moles) in water (150 mL). A white precipitate of antimony silicate forms rapidly. The precipitate is collected by filtration, or the supernatant liquid allowed to evaporate in a large evaporating basin in the fume hood. The solid is allowed to dry and is then packaged and labeled for disposal in a secure landfill.

PICRIC ACID

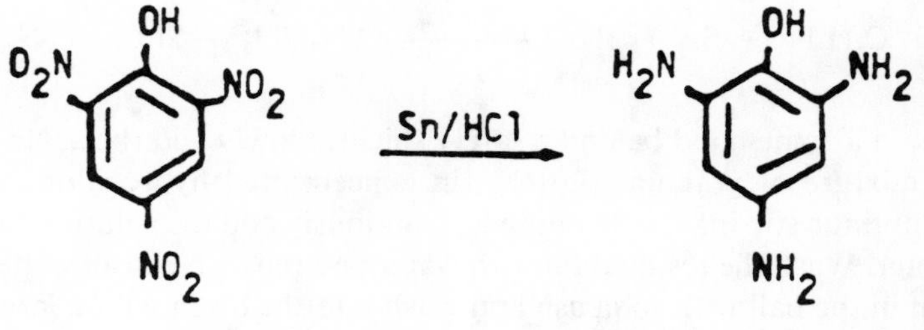

OH, O_2N, NO_2, NO_2 (picric acid) $\xrightarrow{Sn/HCl}$ OH, H_2N, NH_2, NH_2 (triaminophenol)

When dry, solid picric acid is a powerful explosive. In dilute aqueous solution the hazard arises from the possibility of evaporation of the solution on stoppers, caps, or lids of containers, since the explosive can be detonated by friction. Picric acid is smoothly reduced with tin and hydrochloric acid to triaminophenol, which is no longer explosive. It has now been found that very dilute solutions of picric acid are also reduced by acidifying to pH 2 or less with concentrated hydrochloric acid and allowing the solution to stand over granulated tin for 14 days. The color gradually darkens from yellow to brown, and the complete disappearance of the picric acid can be determined by thin layer chromatography of the solution on silica gel plates using methanol:toluene:glacial acetic acid, 8:45:4 as eluant. Picric acid has an R_f

value of about 0.3 and the bright yellow spot is easily visible. The detection limit can be increased by developing the plate in iodine vapor.

AZIDES

Oxidation of inorganic azides by ceric ammonium nitrate is a useful method of disposal. However, for organic azides, this is a slow and unsatisfactory process. Preferable is a reduction with tin and hydrochloric acid. Details of the two procedures are given below.

Inorganic Azides

Work in the fume hood and behind a safety shield. The concentration of the azide solution should be about 1 g/100 mL water. Slowly add 5.5% ceric ammonium nitrate solution (four times the volume of the azide solution) and stir for 1 hour. The solution should show the orange color of ceric ammonium nitrate. Wash the solution into the drain with an excess of water.

Organic Azides

$$C_6H_5N_3 + Sn + 2HCl \longrightarrow C_6H_5NH_2 + SnCl_2 + N_2$$

Work in a fume hood behind a safety shield. Slowly add the azide (1 g) to a stirred mixture of granular tin (6 g) in concentrated hydrochloric acid (100 mL). Continue stirring for 30 minutes. Cautiously add the solution to a pail of cold water. Wash the residual tin with water and reuse. Neutralize the aqueous solution in the pail with soda ash and wash into the drain with a large volume of water.

METAL CARBONYLS

Metal carbonyls such as iron pentacarbonyl and nickel carbonyl are highly toxic and reactive materials and the latter is a suspect carcinogen. These compounds are destroyed by stirring solutions in the appropriate solvent with bleach. The choice of solvent is very important and these are summarized together with quantities and reaction conditions in Table I.

Table I. Oxidation of Metal Carbonyls

	$Fe(CO)_5$ + NaOCl	⟶ stirred for	$Fe(OH)_3$
iron pentacarbonyl solution in hexane:5 mL in 200 mL hexane	bleach 65 mL	30 min under helium	solid filtered, discard aqueous layer to drain, hexane layer recycled or incinerated
	$Fe_2(CO)_9$ + NaOCl	⟶ stirred for	$Fe(OH)_3$
diiron noncarbonyl solution in toluene: 1 g in 100 mL toluene	bleach 50 mL	24 h	solid filtered, discard aqueous layer to drain, toluene layer recycled or incinerated
	$Cr(CO)_6$ + NaOCl	⟶ stirred for	$Cr(OH)_3$
chromium hexacarbonyl solution in tetrahydrofuran: 1 g in 200 mL of THF	bleach 30 mL	15 min	solid filtered, packaged for disposal, aqueous layer to drain. liquid incinerated
	$Ni(CO)_4$ + NaOCl	⟶ stirred for	$Ni(OH)_2$
nickel carbonyl solution in tetrahydrofuran; 5 g in 200 mL of THF	bleach 250 mL	2 h under helium	solid filtered, packaged for disposal. liquid incinerated

CHEMICAL METHODS FOR THE DISPOSAL OF CARCINOGENIC AROMATIC AMINES

Disposal of 4-Aminobiphenyl

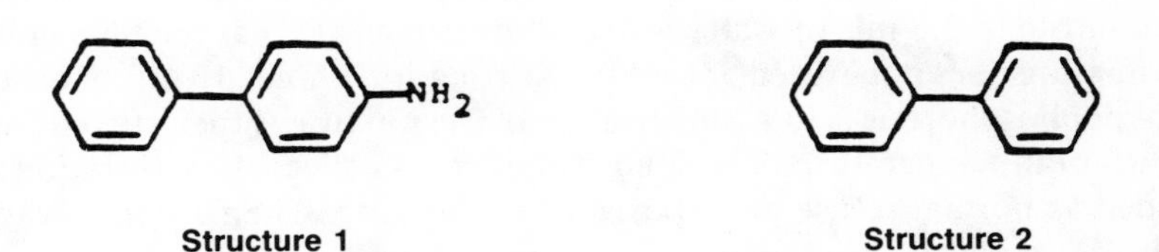

Structure 1 **Structure 2**

Inhalation or absorption through the skin of the dust of 4-aminobiphenyl (Structure 1) has been recognized as a cause of bladder tumors. Removal of the amino group results in the formation of biphenyl (Structure 2), which is not physiologically active. This is accomplished by treating 4-aminobiphenyl with nitrous acid to form a diazonium salt, followed by reduction to biphenyl with hypophosphorous acid. A procedure has been tested in the laboratory and found to result in quantitative conversion of the starting material to product.

To 1.0 g of 4-aminobiphenyl in a 125 mL Erlenmeyer flask is added a mixture of 0.8 mL of water and 2.5 mL of concentrated hydrochloric acid. The mixture is stirred for 10–15 minutes until a homogeneous slurry is formed. The slurry is cooled to 0°C in an ice-salt bath and dropwise, a solution of 1.0 g of sodium nitrite in 2.5 mL of water is added at such a rate that the temperature of the mixture does not rise above 5°C. After stirring for 1 hour, 13 mL of ice-cold 50% hypophosphorous acid is slowly added. Some foaming may occur. After addition of the acid is complete, the mixture is stirred for 18 hours at room temperature. The solid is collected by filtration, the filtrate washed into the drain with a large volume of water, and the solid (biphenyl) discarded with normal refuse or disposed of by burning.

Disposal of 2-Aminofluorene

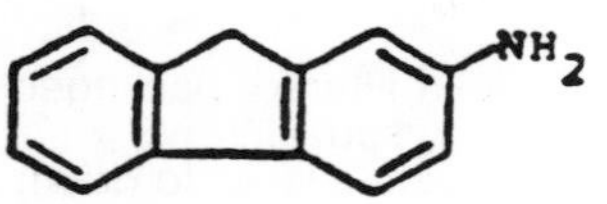

NH_2

Structure 3

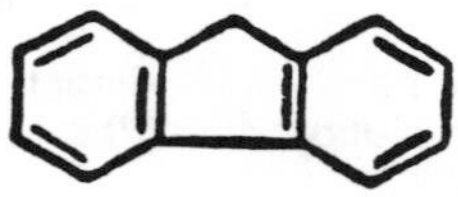

Structure 4

2-Aminofluorene (Structure 3) has been shown to cause cancer in rodents. As in the case of 4-aminobiphenyl (Structure 1), the amino group can be removed by nitrosation and reduction. The solid hydrocarbon fluorene (Structure 4) that is formed can be safely discarded with normal refuse. The conditions for complete conversion of 2-aminofluorene to fluorene have been worked out.

To 1.0 g of 2-aminofluorene in a 125-mL Erlenmeyer flask is added a mixture of 0.8 mL of water and 2.5 mL of concentrated hydrochloric acid. The mixture is stirred for 10–15 minutes until a homogeneous slurry is formed. The slurry is cooled to 0°C in an ice-salt bath and, dropwise, a solution of 1.0 g of sodium nitrite in 2.5 mL of water is added at such a rate that the temperature of the mixture does not exceed 5°C. After stirring for 1 hour, 13 mL of ice-cold 50% hypophosphorous acid is slowly added. The mixture is then stirred for 18 hours at room temperature. The solid (fluorene) is collected by filtration and discarded as normal refuse or disposed of by burning. The filtrate is washed into the drain with a large volume of water.

Disposal of 2-Acetylaminofluorene

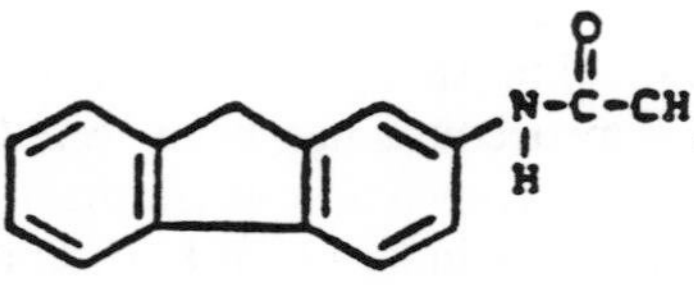

$N(H)-C(=O)-CH_3$

Structure 5

2-Acetylaminofluorene (Structure 5) has been shown to be an animal carcinogen. Like 2-aminofluorene, it can be converted to the nonhazardous compound, fluorene. An extra step is required in the reaction; the acetyl group is removed by acid hydrolysis, and the resulting amino group is removed by nitrosation and reduction. These reactions can be performed sequentially in the same flask.

To 0.25 g of 2-acetylaminofluorene in a 50 mL round-bottomed flask is added 10 mL of 3.6 M hydrochloric acid. The flask is fitted with a condenser and heated at reflux for at least 10 hours, by which time all trace of yellow should have disappeared. The contents of the flask are cooled to 0°C in an ice-salt bath, and, over a period of 5 minutes, a solution of 0.13 g of sodium nitrite in 0.3 mL of water is added dropwise. The mixture is stirred for 30 minutes, and then 2.7 mL of ice-cold 50% hypophosphorous acid is slowly added. After having been stirred at room temperature for 16 hours, the mixture is filtered. The filtrate is washed into the drain with a large volume of water, and the solid (fluorene) discarded with normal refuse or disposed of by burning.

Disposal of N-Hydroxy-2-acetylaminofluorene

$N(OH)-C(=O)-CH_3$

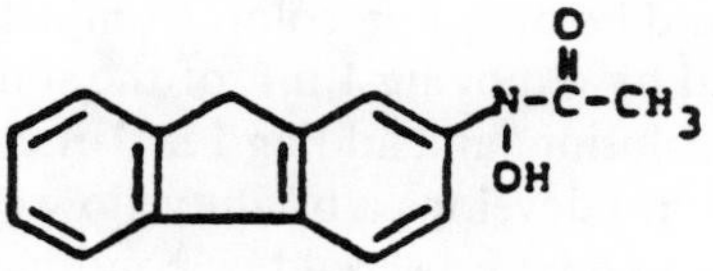

Structure 6

Like 2-acetylaminofluorene, N-hydroxy-2-acetylaminofluorene (Structure 6) is an animal carcinogen which can be hydrolyzed to 2-aminofluorene and then converted to fluorene. Conditions for these conversions are described.

To 0.25 g of N-hydroxy-2-acetylaminofluorene in a 50 mL round-bottomed flask is added 10 mL of 3.6 M hydrochloric acid. The flask is fitted with a condenser and heated under reflux for at least 10 hours. The contents of the flask are cooled to 0°C in an ice-salt bath and, over a period of 5 minutes, a solution of 0.13 g of sodium nitrite in 0.3 mL of water is added dropwise. The mixture is stirred for 30 minutes, and then 2.7 mL of ice-cold 50% hypophosphorous acid is slowly added. After having been stirred at room temperature for 16 hours, the mixture is filtered. The filtrate is washed into the drain with a large volume of water, and the solid (fluorene) discarded with normal refuse or disposed of by burning.

Disposal of 4-Dimethylaminoazobenzene

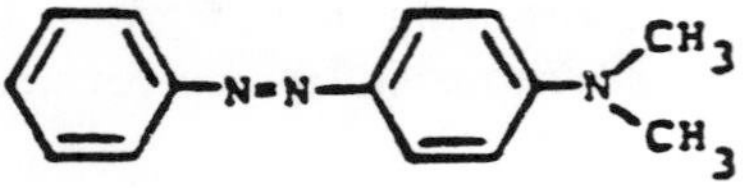

Structure 7 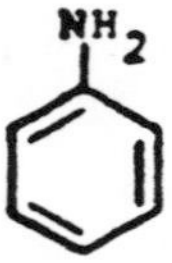 Structure 8 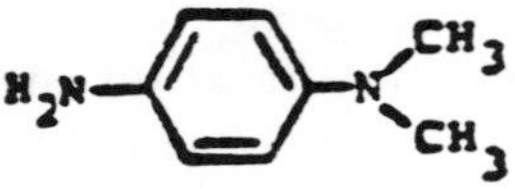 Structure 9

Contact dermatitis has been observed in factory workers using 4-dimethylaminoazobenzene (Structure 7). It also causes cancer in animals. Reduction of the azo group with tin and hydrochloric acid proceeds smoothly to produce aniline (Structure 8) and 4-dimethylaminoanilin (Structure 9). During the checking of the reaction, the products were identified by thin layer chromatography. Both of these compounds are toxic but not carcinogenic. Therefore, they do not present the acute health hazard of 4-dimethylaminoazobenzene. However, they should be disposed of by burning. The procedure we have developed includes a test to ensure that complete conversion of starting material to products has occurred.

To 2 g of 4-dimethylaminoazobenzene in a 250 mL Erlenmeyer flask is added 40 mL of 6 M hydrochloric acid. The deep cherry red mixture is stirred for 15 minutes, then 3.0 g of granulated tin added and stirring continued for 4 hours. The mixture should be brown in color. Complete decomposition of the starting material is tested by removing 1 mL of the solution, neutralizing with 10% sodium hydroxide solution, and adding 1 mL of chloroform. The mixture is shaken. If the chloroform develops a bright yellow color, reaction is incomplete and stirring of the tin mixture should be continued until the test yields a colorless chloroform layer. The reaction mixture is neutralized by the slow addition of 10% sodium hydroxide solution. The mixture is allowed to stand at room temperature for 1 hour, and then the supernatant liquid decanted or the precipitate removed by filtration. The solid is discarded as normal refuse; the liquid is placed in a labeled container for disposal by burning.

Disposal of Urethane

$$NH_2\overset{O}{\overset{\|}{C}}OCH_2CH_3$$

Structure 10

Also known as ethyl carbamate, this urethane (Structure 10) is a cancer suspect agent. The compound can be hydrolyzed in base at room temperature.

To urethane (2 g) contained in a 125 mL Erlenmeyer flask equipped with a magnetic stirrer is added 45 mL of 10% aqueous sodium hydroxide solution. After having been stirred at room temperature for 6 hours or heated under the reflux for 10 minutes, no urethane remains and the liquid can be washed into the drain with a large volume of water.

Disposal of Ethionine

$$CH_3CH_2SCH_2CH_2CH(NH_2)COOH$$

Structure 11

$$CH_3SCH_2CH_2CH(NH_2)COOH$$

Structure 12

The amino acid ethionine (Structure 11) is a cancer suspect agent. All of the physiological properties that it causes can be prevented by the essential amino acid methionine (Structure 12). Ethionine is oxidized and split into fragments by household bleach. The progress of the reaction has been followed by infrared spectroscopy.

Ethionine (1 g) is dissolved in 50 mL of water in a 250 mL Erlenmeyer flask and, with stirring, 60 mL of household bleach (5% sodium hypochlorite) is added. Stirring is continued at room temperature for 2 hours, and then the liquid can be washed into the drain with a large volume of water.

Disposal of Ethidium Bromide

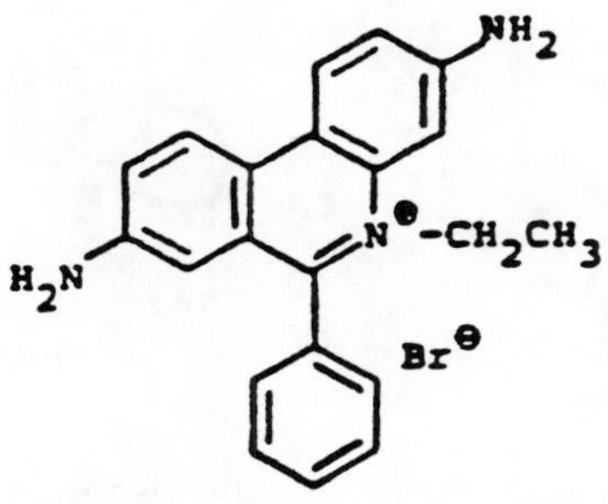

NH_2, H_2N, $N^{\oplus}-CH_2CH_3$, $Br^{\ominus}$

Structure 13

Ethidium bromide (Structure 13) is a compound frequently used in genetic testing, since it intercalates double stranded DNA and RNA. It is known to be mutagenic, and, therefore, must be handled with appropriate precautions.

When a solution of 34 mg of ethidium bromide in 100 mL of water is stirred at room temperature with 300 mL of household bleach for 4 hours, the ethidium bromide is destroyed. The product solution is nonmutagenic in the Ames test, and the major product is 2-carboxybenzophenone (Structure 14), a nontoxic and nonmutagenic material.

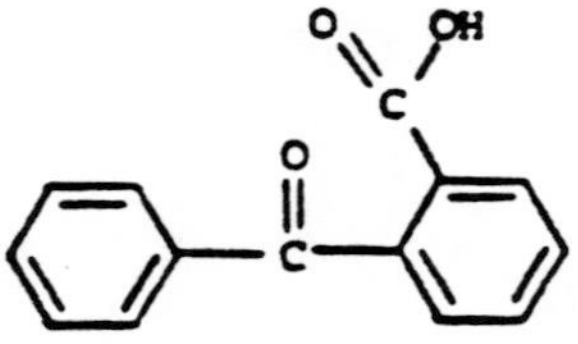

Structure 14

Antineoplastic Agents

Antineoplastic agents are used in chemotherapy, but many of them are themselves carcinogenic. Increasingly, they are being dispensed in hospitals, clinics, and physicians' offices, and there is concern over what to do in case of spills or with waste quantities of the drugs.

The International Agency for Research on Cancer has developed methods for the destruction of seventeen antineoplastic agents, and these have been published.

Two members of the family of N-nitrosoureas that are in use as antineoplastic agents are lomustine (CCNU) (Structure 15) and carmustine (BCNU) (Structure 16).

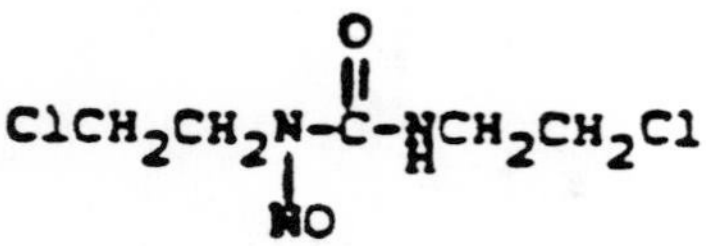

$ClCH_2CH_2N(NO)-C(=O)-NHCH_2CH_2Cl$

Structure 15

$C_6H_{11}-NH-C(=O)-N(NO)-CH_2CH_2Cl$

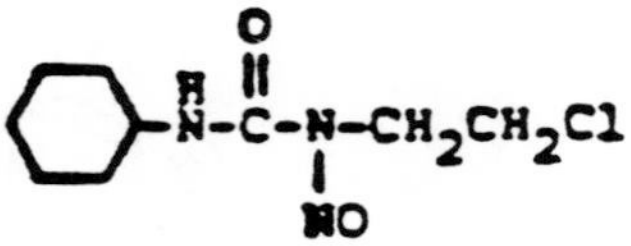

Structure 16

The presently documented method for the disposal of lomustine is treatment with hydrobromic acid in glacial acetic acid, followed by flushing the mixture with nitrogen to remove any nitrosyl bromide formed. There is no known disposal method that converts carmustine to nonmutagenic products.

When lomustine or carmustine (100 mg) are stirred with a solution of 1.2 g of sodium pyrophosphate ($Na_4P_2O_7 \cdot 10H_2O$) in 25 mL of water and 20 drops of Triton X100 at 55°C, both are decomposed in 90 minutes to products that were shown by the Ames test to be nonmutagenic. The major product from BCNU was shown to be N,N'-dicyclohexylurea (Structure 17).

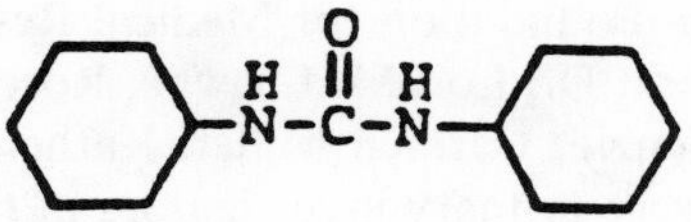

Structure 17

The presence of the surfactant Triton X100 is essential to solubilize the drug. Solutions of the commercially available detergents Sparkleen or liquid Tide were also effective in decomposing these drugs to nonmutagenic residues.

DISPOSAL OF SPILLS OF CHEMICAL CARCINOGENS

Laboratory personnel handling chemical carcinogens should be ready to take appropriate action should a spill occur. Therefore, for each compound that we have developed a waste disposal procedure for, we have also considered a spillage disposal procedure employing the same reagents and conditions as used to destroy surplus material. We recommend that the initial treatment of the spill should depend upon the danger to health that the spilled compound presents. Thus, solid spills of less hazardous compounds are scooped into a container and then treated with the appropriate reagent; spills of solutions of these compounds are absorbed on suitable material and similarly treated. The area of the spill is washed thoroughly with strong soap solution. The 1:1:1 mixture by weight of soda ash, clay cat litter (bentonite), and sand that we found to be useful for reactive chemical spills, has also been found to be excellent as an absorbent for spills of solutions of chemical carcinogens. The absorbed mixture is scooped into a container, transported to the fume hood, and the appropriate chemical reagent added. When reaction is complete, the liquid is washed into the drain with a large volume of water, and the solid discarded as normal refuse.

For compounds such as ethidium bromide that present a considerable health hazard, the appropriate deactivating solution such as bleach, acid, or acidified potassium permanganate is poured directly onto the spill. The liquid is then absorbed using 1:1:1 soda ash, clay cat litter, and sand. This mixture is scooped into a container, transported to the fume hood, and reaction allowed to proceed to completion. The liquid is washed into the drain with a large volume of water, and the solid discarded as normal refuse.

ACKNOWLEDGMENTS

The author gratefully acknowledges financial support from the Heritage Grant Fund of the Occupational Health and Safety Division of Alberta Community and Occupational Health, the Alberta Environment Research Trust, and the Alberta Heritage Foundation for Medical Research.

In addition to the author, Dr. Lois M. Browne, Rosemary Bacovsky, Donna Renecker, Patricia McKenzie, Carmen Miller, Katherine Ayer, John Crerar, and Dr. Roger Klemm were primary investigators in this research project.

BIBLIOGRAPHY

Armour, M. A. "Chemical Waste Management and Disposal," *J. Chem. Educ.*, 65:A64 (1988).

________, R. A. Bacovsky, L. M. Browne, P. A. McKenzie, and D. M. Reneker. *Potentially Carcinogenic Chemicals Information and Disposal Guide* (Edmonton, Canada: University of Alberta, 1986).

________, L. M. Browne, and G. L. Weir. "Tested Disposal Methods for Chemical Wastes from Academic Laboratories," *J. Chem. Educ.* 62:A93 (1985).

________, *Hazardous Chemicals Information and Disposal Guide*, 3rd ed. (Edmonton, Canada: University of Alberta, 1987).

Laboratory Decontamination and Destruction of Carcinogens in Laboratory Wastes: Some Antineoplastic Agents (Lyon, France: International Agency for Research on Cancer, 1985).

Hazardous Chemicals Information and Disposal Guide and *Potentially Carcinogenic Chemicals Information and Disposal Guide* are available from LabStore, 3888 North Fratney Street, Milwaukee, WI 53212 (manager: Ed Segrin, 414-963-8852) or Terochem Laboratories Ltd., P.O. Box 8188, Station F, Edmonton, Alberta T6H 2P1 (telephone 403-438-2222)

CHAPTER 10

The Use Of Reductive and Oxidative Methods To Degrade Hazardous Waste in Academic Laboratories

George Lunn and Eric B. Sansone

INTRODUCTION

Unlike hazardous biological waste, there is no single process (e.g., autoclaving) that can be used to safely degrade hazardous chemical waste. Each chemical or group of chemicals must be considered individually, and a specific degradation procedure devised and validated. However, in recent years, it has become clear that certain procedures are broadly applicable to more than one class of compounds. In this chapter we wish to consider oxidative methods (primarily involving potassium permanganate in sulfuric acid) and reductive methods (primarily involving nickel-aluminum alloy in base). Although many other oxidative or reductive reactions might conceivably be used to degrade hazardous compounds, the bulk of the available literature appears to concentrate on these two methods. In some cases, one method is clearly superior to the other; in other cases, either method may be used. Sometimes neither method works well, and it is necessary to develop other procedures.

This chapter is intended as an overview of some of the published methods for the degradation of hazardous compounds. It cannot be emphasized too strongly that, in all cases, the original literature should be consulted before any degradation reactions are attempted.

NITROSAMINES

When nitrosamines (I) are reduced, two possible products are formed. These are the corresponding amine (II) and the corresponding hydrazine (III) (Scheme I) (Lunn, Sansone, and Keefer, *Carcinogenesis*, 1983). A number of hydrazines have been shown to be carcinogenic (Toth, 1977) and mutagenic (Lunn, unpublished results), whereas the amines are generally felt to be relatively innocuous. A reductive degradation process should produce only the amine with no trace of the hydrazine. An investigation of a number of reduc-

$$\underset{\text{I}}{RR'N{-}NO} \longrightarrow \underset{\text{II}}{RR'NH} + \underset{\text{III}}{RR'N{-}NH_2}$$

Scheme I

ing systems (Lunn, Sansone, and Keefer, *Carcinogenesis*, 1983) showed that only nickel-aluminum alloy in dilute base completely reduced the nitrosamines. It produced the corresponding amines with no trace of the hydrazines (although hydrazines were detected as transient intermediates). It was also found that the final reaction mixtures were nonmutagenic (Lunn, unpublished results). A collaborative study organized by the International Agency for Research on Cancer (*Some* N-*Nitrosamines*, IARC, 1982) found that the method worked well, and that it was the method of choice for the destruction of nitrosamines in water, ethanol, olive oil, and petri dishes. It was also found that the method could be used for the destruction of undiluted nitrosamines and the treatment of solutions of nitrosamines in dichloromethane. The collaborative study also found that potassium permanganate in sulfuric acid (Castegnaro, Michelon, and Walker, 1982) could be used to degrade nitrosamines, and that this procedure was the method of choice for the treatment of spills of aqueous solutions of nitrosamines where the homogeneous nature of the reaction is a great advantage. The procedure was also recommended for the treatment of bulk quantities and aqueous solutions of nitrosamines. Thus, either oxidative or reductive methods may be used to degrade nitrosamines, and the choice generally depends on the matrix in which the nitrosamine is found.

Other methods that the IARC study found to be applicable to the destruction of nitrosamines were reaction with hydrogen bromide in glacial acetic acid and reaction with triethyloxonium salts.

HYDRAZINES

Hydrazines are also readily degraded by using nickel-aluminum alloy in dilute base (Lunn, Sansone, and Keefer, *Environmental Science and Technology*, 1983). The hydrazines(III) are completely degraded and the only products are the corresponding amines(II). In a collaborative study organized by the IARC (*Some Hydrazines*, 1983), this procedure was the method of choice for the treatment of undiluted hydrazines and solutions in water, methanol, oil, and dimethyl sulfoxide (DMSO). It was also recommended for the treatment of petri dishes contaminated with hydrazines. Potassium permanganate in sulfuric acid also degraded hydrazines, and this procedure was the method of choice for treating solutions of hydrazines in ethanol and in organic solvents

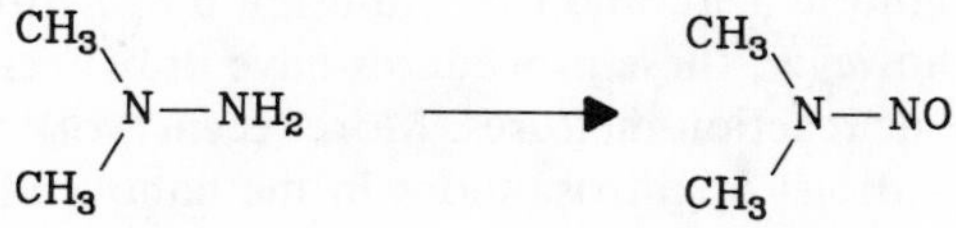

$$(CH_3)_2N-NH_2 \longrightarrow (CH_3)_2N-NO$$

IV V

Scheme II

not miscible with water, and for decontaminating glassware. The method could also be used to treat undiluted hydrazines and solutions in water, oil, and methanol, and to treat spills. Two other oxidative methods were tested. One was the use of potassium iodate, which was the method of choice for treating spills, and could also be used to treat undiluted hydrazines and solutions in water, methanol, ethanol, and organic solvents not miscible with water. The other oxidative method was the use of hypochlorites (either sodium hypochlorite or calcium hypochlorite), which was the method of choice for treating laboratory equipment, protective clothing, and animal litter. It could also be used to treat undiluted hydrazines, solutions in water, methanol, and ethanol, and to decontaminate glassware and petri dishes.

Later work (Castegnaro et al., 1986), however, showed that, although these oxidative procedures did indeed completely degrade the hydrazines, in a number of cases carcinogenic nitrosamines were produced. 1,1-Dimethylhydrazine (IV) was particularly prone to undergo this reaction, being readily oxidized to *N*-nitrosodimethylamine (V) (Scheme II). Nitrosamines were also found in the products from the degradation reactions of other hydrazines, and, in a number of instances, the final reaction mixtures were mutagenic. For these reasons, it was concluded that bulk quantities and solutions of hydrazines should only be degraded using the nickel-aluminum alloy procedure, and that the oxidative methods should only be used for the treatment of spills and the decontamination of glassware where the heterogeneous nature of the nickel-aluminum alloy reaction renders it inappropriate.

NITROSAMIDES

A collaborative study organized by the IARC investigated a number of methods for the degradation of nitrosamides and found (*Some N-Nitrosamides*, IARC, 1983) that potassium permanganate in sulfuric acid was the method of choice for treating spills and decontaminating glassware, and that the procedure could also be used to degrade bulk quantities of nitrosamides and to treat solutions in methanol, ethanol, acetone, water, and volatile organic solvents. Other methods that this study found useful for degrading nitrosamides were the use of hydrogen bromide in glacial acetic acid, sulfamic acid in 6 M hydrochloric acid, and iron filings in 6 M hydrochloric acid. In a number of cases, however, these procedures have drawbacks such as the production of mutagenic reaction mixtures. More recent work (Lunn et al., 1988) has shown that solutions of nitrosamides in methanol, ethanol, acetone, or DMSO can be completely decontaminated by adding nickel-aluminum alloy, and then slowly raising the pH of the solution by the sequential addition of sodium bicarbonate, sodium carbonate, and, finally, potassium hydroxide solution. In all cases, the nitrosamides are completely destroyed, and the final reaction mixtures are nonmutagenic. The products that can be identified are also nonmutagenic. We hypothesize that the slowly increasing pH causes degradation of the nitrosamides at a slow enough rate that any toxic and explosive diazoalkanes generated can react with the solvent. In any event, none are detected. This is in contrast to the case where strong base is used when large quantities are found. The reducing action of the nickel-aluminum alloy appears necessary for the suppression of mutagens. When the alloy is omitted, mutagenic activity is found in a number of instances.

RELATED COMPOUNDS CONTAINING N-N AND N-O BONDS

The use of nickel-aluminum alloy in dilute base is a general procedure for the cleavage of N-N and N-O bonds (Lunn, Sansone, and Keefer, 1985). As well as its use in degrading nitrosamines and hydrazines, it can also be used to reduce hydroxylamines, nitramines, *N*-oxides, tetrazenes, triazenes, and azo- and azoxy-compounds to their parent amines. In devising a degradation reaction, the nature of the product should always be borne in mind, but, in many cases, the nickel-aluminum alloy reaction can be used to effect complete degradation and produce only nontoxic, or at least significantly less toxic, products.

Nickel-aluminum alloy (Lunn and Sansone, *American Journal of Hospital Pharmacy*, 1987) has also been shown to degrade the drugs dacarbazine, procarbazine, isoniazid, and iproniazid. The last three are hydrazines and dacarbazine is a triazene. Dacarbazine (VI), which is mutagenic, was reduced to the nonmutagenic amine (VII) and dimethylamine (Scheme III). When potassium permanganate in sulfuric acid was used to treat dacarbazine, the drug was completely degraded, but a small amount of the potent carcinogen *N*-nitrosodimethylamine (V) was detected and the final reaction mixture was mutagenic. By greatly increasing the oxidant:dacarbazine ratio, nonmutagenic reaction mixtures could be produced. In this case, however, the reaction volumes became inconveniently large. Thus, in this instance, the nickel-aluminum alloy method (which produced only nonmutagenic reaction mixtures) is clearly preferable to potassium permanganate.

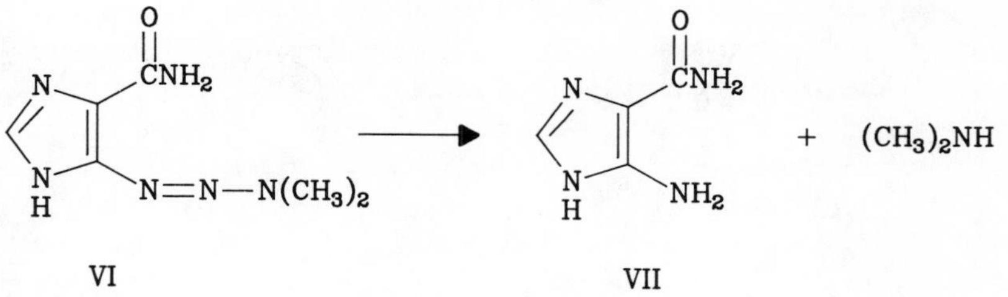

VI → VII + $(CH_3)_2NH$

Scheme III

ANTINEOPLASTIC DRUGS

A collaborative study organized by the IARC (*Some Antineoplastic Agents*, 1985) found that the antineoplastic drug streptozotocin (which is an *N*-nitroso compound) could be completely degraded by using potassium permanganate in sulfuric acid. Only nonmutagenic reaction mixtures were produced. On the other hand, the same study found that, although the procedure completely degraded the related *N*-nitroso antineoplastics carmustine (BCNU), lomustine (CCNU), semustine (MeCCNU), PCNU, and chlorozotocin, the final reaction mixtures showed high mutagenic activity, and the procedure could not be recommended for these drugs. Streptozotocin could also be degraded using hydrogen bromide in glacial acetic acid.

Other drugs that could be degraded using potassium permanganate in sulfuric acid were doxorubicin, daunorubicin, methotrexate, dichloromethotrexate, vincristine, vinblastine, 6-thioguanine and 6-mercaptopurine. Methotrexate could also be degraded using sodium hypochlorite.

Since the conclusion of the collaborative study, it has been found that nickel-aluminum alloy in dilute base can be used to degrade the *N*-nitroso antineoplastic drugs BCNU, CCNU, MeCCNU, PCNU, and chlorozotocin as well as the related antineoplastic drugs mechlorethamine, melphalan, chlorambucil, cyclophosphamide, ifosfamide, uracil mustard, and spirohydantoin mustard that contain 2-chloroethylamino groups (Lunn et al., 1989). In each case, the drug was completely destroyed and nonmutagenic reaction mixtures were produced.

POLYCYCLIC AROMATIC HYDROCARBONS

A study organized by the IARC (*Some Polycyclic Aromatic Hydrocarbons*, 1983) found that three methods were effective for the degradation of polycyclic aromatic hydrocarbons (PAH). These methods were the use of potassium permanganate in sulfuric acid, saturated aqueous potassium permanganate, and concentrated sulfuric acid. The method of choice in any given situation depended on the PAH to be degraded and the matrix in which it was found.

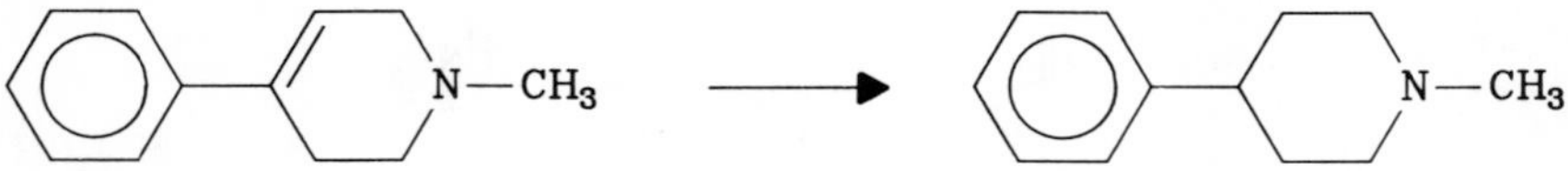

N—CH_3 → N—CH_3

VIII IX

Scheme IV

In a small feasibility study (Lunn, unpublished results), it was found that phenanthrene could be completely degraded with nickel-aluminum alloy in dilute base. However, the reaction produced a variety of partially hydrogenated phenanthrenes, not the fully saturated compound. Since a number of partially hydrogenated PAH's are known (Lijinsky, Garcia, and Saffiotti, 1970) to be more toxic than their parent compounds, it appears that potassium permanganate is clearly superior to nickel-aluminum alloy for degrading PAH's.

AROMATIC AMINES

Clearly nickel-aluminum alloy in dilute base is likely to have little effect on aromatic amines. A collaborative study organized by the IARC (*Some Aromatic Amines and 4-Nitrobiphenyl*, 1985) found that potassium permanganate in sulfuric acid was the method of choice for the treatment of bulk quantities, solutions in water, oil, and volatile organic solvents, and for the decontamination of spills and glassware. The method could also be applied to the degradation of 4-nitrobiphenyl after it was first reduced to 4-aminobiphenyl with zinc and dilute sulfuric acid. Other methods recommended by the IARC were treatment with horseradish peroxidase/hydrogen peroxide, and diazotization followed by either hypophosphorous acid or heat treatment.

1-METHYL-4-PHENYL-1,2,3,6-TETRAHYDROPYRIDINE

1-Methyl-4-phenyl-1,2,3,6-tetrahydropyridine (MPTP) (VIII) is a potent neurotoxin that produces the symptoms of Parkinson's disease in humans. It is used in some biomedical research laboratories. It was found (Pitts et al., 1986; Yang et al., 1988) that bulk quantities of MPTP and MPTP dissolved in water, methanol, ethanol, DMSO, acetone, and water/acetonitrile high-pressure liquid chromatography (HPLC) eluant could be completely degraded by using potassium permanganate in sulfuric acid. Nickel-aluminum alloy in dilute base could be used to reduce MPTP to the pharmacologically inactive (Cohen and Mytilineou, 1985) 1-methyl-4-phenylpiperidine (IX) (Scheme IV), but the procedure tended to produce variable results and could not be recommended as a destruction procedure. In this case, potassium permanganate is clearly superior to nickel-aluminum alloy.

CYANOGEN BROMIDE

Cyanogen bromide and sodium cyanide may be safely destroyed by basification with sodium hydroxide and oxidation with sodium or calcium hypochlorite (Lunn and Sansone, *Analytical Biochemistry*, 1985). In view of the ease of the reaction, reductive processes were not investigated.

AFLATOXINS

Aflatoxins, which are carcinogenic fungal metabolites, may be successfully degraded using sodium hypochlorite in the presence of acetone or by using potassium permanganate in the presence of sulfuric acid (IARC, *Aflatoxins*, 1980). Under certain circumstances, ammonia and quicklime can also be used to effect destruction of aflatoxins.

OTHER COMPOUNDS

Ethidium bromide (EB) is a powerful mutagen widely used in biomedical laboratories for visualizing DNA fragments. Nickel-aluminum alloy in dilute base, potassium permanganate in sulfuric acid, and sodium hypochlorite all degraded EB (Lunn and Sansone, *Analytical Biochemistry*, 1987), but they also produced mutagenic reaction mixtures. It was found that sodium nitrite in the presence of hypophosphorous acid caused complete degradation of the EB and produced only nonmutagenic reaction mixtures, although the exact course of the reaction is uncertain. Thus, in this case, neither oxidative nor reductive processes work for the compound in question, and we must try a different reaction. There has been some dispute as to whether or not sodium hypochlorite produces mutagenic reaction mixtures when used to degrade ethidium bromide, and, accordingly, we have included the raw data used to support Lunn and Sansone's conclusions (*Analytical Biochemistry*, 1987) as Table I. Recently, Quillardet and Hofnung (1988) have confirmed that the sodium hypochlorite degradation of ethidium bromide does produce mutagenic reaction mixtures. These authors also report that degradation of high (10 mg/mL) concentrations of ethidium bromide using the sodium nitrite/hypophosphorous acid method results in slightly mutagenic reaction mixtures. This method has only been validated (Lunn and Sansone, *Analytical Biochemistry*, 1987), however, for somewhat lower concentrations (0.5 mg/mL), and so, before decontamination, the solutions should either be diluted to this concentration, or increased quantities of sodium nitrite and hypophosphorous

Table I. Destruction of Ethidium Bromide: Number of Revertants Found When Final Reaction Mixtures are Tested

		Strain and Activation									
		TA1530		TA1535		TA100		TA1538		TA98	
		O	S9	O	S9	O	S9	O	S9	O	S9
Method[a]	Buffer										
A	Water	25	12	19	26	140	148	16	39	*79*	41
	MOPS	40	*45*	32	*38*	170	140	14	34	*68*	54
	TBE	16	14	19	16	216	158	11	43	16	*83*
	CsCl	38	12	14	22	120	136	20	32	51	36
B	Water	24	8	20	16	179	143	16	36	*69*	38
	MOPS	26	15	22	23	148	135	16	35	47	44
	TBE	22	12	13	17	178	147	18	36	*127*	36
	CsCl	18	10	17	21	207	170	19	43	*124*	64
C	Water	20	12	30	33	142	183	5	32	33	47
	MOPS	30	14	26	24	175	152	18	39	*107*	*80*
	TBE	25	15	28	28	198	164	*39*	26	*114*	18
	CsCl	39	20	22	14	192	151	34	*53*	*92*	46
D	Water	45	23	*39*	10	180	157	14	22	26	45
	MOPS	47	31	22	4	131	177	13	38	32	38
	TBE	47	23	*75*	*43*	*330*	*421*	*40*	35	*54*	60
	CsCl	*132*	*100*	*117*	*80*	*609*	*276*	19	42	49	78
E	Water	20	20	19	16	215	189	27	26	*115*	65
	MOPS	17	30	27	27	203	*261*	*68*	*124*	*145*	*177*
	TBE	21	16	20	27	190	186	*108*	*754*	*174*	*696*
	CsCl	35	24	26	17	166	155	*36*	44	*84*	58
F	Water	*102*	34	18	18	27	234	*182*	*60*	43	54
	MOPS	*64*	*35*	*47*	*33*	*682*	*566*	*49*	*54*	*174*	101
	TBE	6	5	4	12	155	*406*	3	*174*	4	*262*
	CsCl	26	17	19	19	194	209	*38*	*61*	*92*	*81*
G	Water	14	8	24	26	100	121	16	*79*	16	*127*
	MOPS	18	6	9	16	99	127	14	*103*	38	*189*
	TBE	4	2	21	8	85	118	4	*174*	19	*227*
	CsCl	18	10	12	13	95	124	10	*107*	25	*127*
H	Water	30	20	10	18	113	104	18	22	27	51
	MOPS	15	28	26	23	115	112	14	36	33	42
	TBE	31	20	24	22	136	95	8	32	29	45
	CsCl	28	14	11	14	127	122	30	32	42	36
Original Solution[b]	Water	14	4	20	18	156	190	18	*2523*	20	*3480*
	MOPS	23	15	12	20	177	200	28	*2291*	32	*2219*
	TBE	21	16	21	25	144	182	24	*2552*	30	*3015*
	CsCl	40	13	19	14	130	158	25	*2030*	37	*2610*
Control	Water	25	12	16	18	130	170	17	22	22	49
	Buffer	29	16	17	15	124	125	17	23	26	39

Note: The above figures are the supporting data for Lunn and Sansone (*Analytical Biochemistry*, 1987). Procedures are as specified in that paper. Values were significantly mutagenic (shown in italics) when the number of revertants were more than twice background. Note that the various figures may not be strictly comparable because each degradation procedure produces various degrees of dilution. For example, Method H would have a volume increase from 3.00 mL to 3.96 mL.

[a] Method A Potassium permanganate
Method B Potassium permanganate (greater amount of oxidant)

Table I continued

Method C	Potassium permanganate/sulfuric acid
Method D	Potassium permanganate/sulfuric acid (greater amount of oxidant)
Method E	Sodium hypochlorite
Method F	Sodium hypochlorite (greater amount of oxidant)
Method G	Nickel-aluminum alloy/sodium hydroxide solution
Method H	Sodium nitrite/hypophosphorous acid

[b]The concentrations of ethidium bromide in the original solutions were: Water, 0.247 mg/mL; MOPS, 0.253 mg/mL; TBE, 0.250 mg/mL; and CsCl, 0.302 mg/mL.

acid should be used. Quillardet and Hofnung (1988) also state that potassium permanganate in hydrochloric acid can be used to degrade ethidium bromide in solution. In our hands, however, this method gave some mutagenic reaction mixtures. Our data are included as Tables II and III. These results are unusual in that the reaction mixtures were markedly more mutagenic when they were

Table II. Degradation of Ethidium Bromide Solutions by Potassium Permanganate/Hydrochloric Acid *Without* Stirring

		Strain and Activation									
		TA1530		TA1535		TA98		TA100		TA1538	
Initial Solvent	Concentration (mg/mL)	0	S9	0	S9	0	S9	0	S9	0	S9
TRIS	10.080	20	17	16	20	*75*	66	194	171	11	38
	0.504	25	14	24	21	20	40	161	144	12	20
	0	26	12	14	18	18	27	165	147	14	15
MOPS	10.000	24	26	18	30	48	57	230	178	16	32
	0.493	*95*	30	*49*	42	26	41	206	185	16	36
	0	*72*	*66*	41	45	29	28	215	186	7	24
H_2O	10.000	11	17	17	12	*106*	62	217	126	18	34
	0.500	18	18	31	12	34	65	151	170	17	50
	0	21	14	19	16	41	59	116	160	15	48
CsCl	0.953	30	18	21	19	36	58	136	148	16	21
	0.477	32	6	20	22	43	61	126	127	*32*	43
	0	35	14	15	20	36	59	134	165	18	35
Control Values		30	15	23	22	27	48	154	160	12	27

Note: Reactions are based on the method of Quillardet and Hofnung (1988). A solution of ethidium bromide in an appropriate solvent (3 mL) was mixed with 0.5 M potassium permanganate solution (3 mL) and then 2.5 M hydrochloric acid (3 mL) was added. The mixture was allowed to stand at room temperature for 3 hours, at which time no potassium permanganate was visible. The reaction mixtures were neutralized with 0.8 mL 10 M sodium hydroxide solution, and then 0.5 mL glacial acetic acid was added. In some cases, the purple color of potassium permanganate was now visible, and the reaction mixtures were decolorized with sodium ascorbate. Fluorescence testing showed no trace of ethidium bromide (<0.0005 mg/mL) in all cases. High concentrations of ethidium bromide in CsCl solution could not be obtained. The concentrations used are the best that were obtainable at the time. None of the buffers were mutagenic. Solutions containing high (10 mg/mL) concentrations of ethidium bromide were tested for mutagenicity, but were found to be toxic to the bacteria. Details of buffer solutions, fluorescence assays, and mutagenesis assays were as before. (Lunn and Sansone, *Analytical Biochemistry*, 1985).

Table III. Degradation of Ethidium Bromide Solutions by Potassium Permanganate/Hydrochloric Acid *With* Stirring

		Strain and Activation									
		TA1530		TA1535		TA98		TA100		TA1538	
		0	S9	0	S9	0	S9	0	S9	0	S9
Initial Solvent	Concentration (mg/mL)										
TRIS	10.077	29	12	21	29	37	31	150	166	15	18
	0.503	*89*	*50*	34	*41*	24	44	162	156	11	11
	0	*221*	*145*	*80*	*74*	29	47	190	198	11	30
MOPS	10.047	*236*	*101*	*96*	*72*	*73*	50	*566*	269	22	18
	0.503	*580*	*100*	*155*	*65*	110	64	*479*	*392*	TX	17
	0	*146*	*61*	*77*	*49*	49	69	254	209	18	23
H_20	10.027	*77*	30	*72*	*39*	*126*	62	*508*	251	*44*	18
	0.500	18	15	15	15	36	39	140	105	11	30
	0	28	12	18	30	37	37	134	142	19	12
CsCl	0.638	27	10	17	19	46	47	132	159	18	29
	0	21	8	16	19	42	75	123	140	16	*45*
Control Values		21	17	20	16	32	58	130	155	16	21

Note: Reactions are based on the method of Quillardet and Hofnung (1988). A solution of ethidium bromide in an appropriate solvent (3 mL) was mixed with 0.5 M potassium permanganate solution (3 mL), and then 2.5 M hydrochloric acid (3 mL) was added. The mixture was stirred at room temperature for 3 hours, at which time no potassium permanganate was visible except for the CsCl reactions, the 0.500 mg/mL water solution, and the water blank. These solutions were decolorized with sodium ascorbate after neutralization. The reaction mixtures were neutralized with 0.8 mL 10 M sodium hydroxide solution, and then 0.5 mL glacial acetic acid was added. Fluorescence testing showed no trace of ethidium bromide (<0.0005 mg/mL) in all cases. High concentrations of ethidium bromide in CsCl solution could not be obtained. The concentration used was the best obtainable at the time. None of the buffers were mutagenic. Details of buffer solutions, and fluorescence assays, and mutagenesis assays were as before. (Lunn and Sansone, *Analytical Biochemistry*, 1987).

stirred than when they were not stirred. Also, the incidence of mutagenicity did sometimes increase as the concentration of ethidium bromide *decreased*. In any event, based on these results, it would be difficult to recommend this reaction as a decontamination procedure.

Other compounds, for which validated destruction procedures that do not involve reductive or oxidative processes have been published, include dimethyl sulfate (Lunn and Sansone, *American Industrial Hygiene Association Journal*, 1985) and cyclophosphamide, ifosfamide, cisplatin, CCNU, and chlorozotocin (IARC, *Some Antineoplastic Agents*, 1985).

OTHER REVIEWS

The destruction of chemical carcinogens (Lunn, Castegnaro, and Sansone, 1988) and the use of nickel-aluminum alloy as a reducing agent (Keefer and Lunn, 1989) have recently been reviewed.

ACKNOWLEDGMENT

Research for this chapter was supported by a grant from the National Cancer Institute, DHHS, under contract NO1-CO-74102 with Program Resources, Inc.

BIBLIOGRAPHY

Castegnaro, M., J. Michelon, and E. A. Walker. In: Bartsch, H., I. K. O'Neill, M. Castegnaro, and M. Okada, Eds. N-*Nitroso Compounds: Occurrence and Biological Effects* (Lyon, France: International Agency for Research on Cancer, 1982, p. 151).

Castegnaro, M., I. Brouet, J. Michelon, G. Lunn, and E. B. Sansone. *American Industrial Hygiene Association Journal* 47:360 (1986).

Cohen, C., and C. Mytilineou. *Life Sciences* 35:237 (1985).

Keefer, L. K., and G. Lunn. *Chemical Reviews* 89:459 (1989).

Laboratory Decontamination and Destruction of Aflatoxins B_1, B_2, G_1, G_2 in Laboratory Wastes. (Lyon, France: International Agency for Research on Cancer, 1980).

Laboratory Decontamination and Destruction of Carcinogens in Laboratory Wastes: Some N-*Nitrosamines* (Lyon, France: International Agency for Research on Cancer, 1982).

Laboratory Decontamination and Destruction of Carcinogens in Laboratory Wastes: Some Hydrazines (Lyon, France: International Agency for Research on Cancer, 1983).

Laboratory Decontamination and Destruction of Carcinogens in Laboratory Wastes: Some N-*Nitrosamides* (Lyon, France: International Agency for Research on Cancer, 1983).

Laboratory Decontamination and Destruction of Carcinogens in Laboratory Wastes: Some Polycyclic Aromatic Hydrocarbons. (Lyon, France: International Agency for Research on Cancer, 1983).

Laboratory Decontamination and Destruction of Carcinogens in Laboratory Wastes: Some Antineoplastic Agents (Lyon, France: International Agency for Research on Cancer, 1985).

Laboratory Decontamination and Destruction of Carcinogens in Laboratory Wastes: Some Aromatic Amines and 4-Nitrobiphenyl (Lyon, France: International Agency for Research on Cancer, 1985).

Lijinsky, W., H. Garcia, and U. Saffiotti. *Journal of the National Cancer Institute* 44:641 (1970).

Lunn, G., M. Castegnaro, and E. B. Sansone. In: Woo, Y.-t., D.Y. Lai, J. C. Arcos, and M. F. Argus, Eds. *Chemical Induction of Cancer*, Vol. IIIC (San Diego: Academic Press, 1988, p. 171).

Lunn, G., and E. B. Sansone. *American Industrial Hygiene Association Journal* 46:111 (1985).

________. *Analytical Biochemistry* 147:245 (1985).

________. *American Journal of Hospital Pharmacy* 44:2519 (1987).

________. *Analytical Biochemistry* 162:453 (1987).

Lunn, G., E. B. Sansone, A. W. Andrews, and L. K. Keefer. *Cancer Research* 48:522 (1988).

Lunn, G., E. B. Sansone, A. W. Andrews, and L. C. Hellwig. *Journal of Pharmaceutical Sciences* 78:652 (1989).

Lunn, G., E. B. Sansone, and L. K. Keefer. *Carcinogenesis* 4: 315 (1983).

________. *Environmental Science and Technology* 17:240 (1983).

________. *Synthesis* 1104 (1985).

Pitts, S. M., S. P. Markey, D. L. Murphy, A. Weisz, and G. Lunn. In: Markey, S. P., N. Castagnoli, Jr., A. Trevor, and I. Kopin, Eds. *MPTP: A Neurotoxin Producing a Parkinsonian Syndrome* (Orlando, FL: Academic Press, 1986, p. 703).

Quillardet, P., and M. Hofnung. *Trends in Genetics* 4:89 (1988).

Toth, B. *Cancer* 40:2477 (1977).

Yang, S. C., S. P. Markey, K. S. Bankiewicz, W. T. London, and G. Lunn. *Laboratory Animal Science* 38:563 (1988).

CHAPTER 11

Chemical/Physical Methods for Treatment and Disposal of University Hazardous Wastes

Sharon Ward Harless

INTRODUCTION

Colleges and universities generate chemical hazardous wastes that must be managed in accordance with regulations enacted by the U.S. Environmental Protection Agency (EPA). The EPA regulations were designed and written for industry where very large volumes of only a few wastes are the norm. In contrast, universities generate very small volumes of literally thousands of different chemical wastes. For example, in 1987 Iowa State University (ISU) generated over 3000 different chemical wastes, but had a total volume of only 35,000 kg.

At one time, ISU used to dispose of all these wastes by the traditional method of "lab packing" in 55 gal drums and shipping the drums to an EPA permitted land disposal facility. This is tedious work, very time consuming, and has proven to be quite expensive. In addition, it presents a perpetual liability problem because of EPA's "cradle-to-grave" regulations. Several universities have experienced unexpectedly high costs when a waste handler or land disposal facility has gone out of business.

In view of the cost of a lab pack and landfill hazardous waste disposal policy, the Environmental Health and Safety department at ISU conducted an inventory of the university's hazardous waste streams to identify chemical wastes that would lend themselves to local treatment, thereby eliminating landfill disposal of these hazardous chemicals.

IMPLEMENTING CHEMICAL/PHYSICAL TREATMENT METHODS

After the inventory, those of us in the Environmental Health and Safety Department conducted a search of several disciplines and sources of literature to determine treatment methods we could use. Those methods we discovered were then tested and modified to make them as simple as possible.

Most of the treatment methods we finally adopted do not require sophisticated equipment beyond that available in a normal laboratory. They are simple chemical routines that can be carried out by someone with two years of college chemistry. Some of the treatment methods can be carried out in either a small laboratory, or scaled to whatever capacity is needed. Our department collects the waste chemicals and conducts the treatments in one building, the Chemical Waste Handling Facility (CWHF).

The implementation of ISU's program required a considerable initial investment in building, equipment, and supplies. Equipment purchased included a large walk-in hood, bench hoods, distillation apparatus, a hydraulic filter press, pumps, storage shelves and containers, and basic laboratory equipment. If a university does not want to commit to this initial investment, treatment methods can still be implemented at the laboratory level.

EPA REGULATORY REQUIREMENTS

The EPA regulates the generation, treatment, storage, and disposal of hazardous waste. If universities are going to investigate the possibilities of utilizing chemical/physical treatment methods to treat or dispose of their hazardous waste, they must understand how the EPA regulations apply to them. The EPA distinguishes between large and small quantity generators in applying regulation requirements for hazardous waste generators. However, the EPA regulations do not make any distinctions in size for treatment, storage, and disposal facilities. Therefore, if a university decides to collect the hazardous waste generated on its campus and employ chemical/physical treatment methods to destroy or detoxify the waste, it will first need to obtain an EPA Waste Treatment, Storage, and Disposal Facility permit.

The EPA permitting process is very complex and arduous. To obtain an operating permit for a new hazardous waste facility takes approximately two years. It is unfortunate that the EPA regulations act as a barrier to the implementation of a hazardous waste treatment program.

There is an exemption to the EPA permitting requirements for chemical/physical treatment of hazardous waste applicable to universities. In order for a university to implement chemical/physical treatment methods without obtaining an EPA permit, chemical waste would have to be treated as it is produced in individual laboratories. As long as the chemical/physical treatment is part of the actual experiment, the waste can be treated without a permit. However, it cannot be collected, stored, and then treated. This would mean that researchers would have to treat all of their wastes themselves; it would not be possible to set up an Environmental Health and Safety Office to come to each lab to treat the wastes. Therefore, a word of warning: Before implementing a chemical/physical treatment program for hazardous waste, research the EPA regulations and talk to Federal and State Agencies that enforce the EPA regulations.

FINANCIAL DETAILS

Treatment Methods and Associated Costs

Although the implementation and operation of a chemical/physical treatment program at ISU required a considerable initial investment, the savings we have had prove it is cost effective. Table I gives a breakdown of the treatment methods used by the Environmental Health and Safety department at ISU. The table lists the amount of hazardous waste generated by ISU in 1987 and the amount it would have cost the university to dispose of this waste off-site at an EPA permitted commercial facility. The unit disposal costs were taken from a study done by Peter Ashbrook of the University of Illinois, published in the American Chemical Society (ACS) *Network News*, 31 July 1987. Table I then lists the dollar cost to ISU for treating this waste on-site and indicates the cost savings.

Investment Costs of Treatment Program

Table II lists the capital investment that was required to implement and operate the chemical/physical treatment program at ISU.

Operating Costs for Treatment Methods

Table III lists the operating costs that were incurred to carry out the treatment processes.

After the capital investment costs, labor costs, and operating costs were included, the total cost savings of the treatment program amounted to a total of $295,930 per year.

DESCRIPTION OF CHEMICAL/PHYSICAL TREATMENT METHODS

Acid/Base Neutralization

This treatment method is simple, requiring only the use of a hood and corrosion-resistant containers and equipment. The person who does this treatment should wear a face shield, rubber apron, rubber boots, neoprene gloves, and a half-face respirator if the volume is large. This process can be done in the laboratory, or the waste can be collected and treated at a treatment facility.

Distillation

The Environmental Health and Safety department distills spent solvent that is generated on campus and sells it back to ISU researchers. Currently, the

Table I. Cost Savings and Comparisons

Treatment Conducted at ISU		List of Costs Which ISU Would Have Faced If Treatment Had Not Been Done			Savings	
Treatment Method	Amount Treated at ISU (Kg/Yr)	Unit Cost of Commercial Disposal (Dollar/Kg)	Total Cost Commercial Disposal	Commercial Disposal Method	ISU Treatment Operating Costs	ISU Net Savings
Acid/Base Neutralization	1,200	$ 20	$ 24,000	Lab Pack Treatment	$ 600	$ 23,400
Bulking Solvents	1,180	$ 20	$ 23,600	Lab Pack Incineration	$ 500	$ 23,100
Distillation	5,400	$ 1	$ 5,400	Bulk Incineration	$ 250	$ 5,150
Incineration - Vet Med	5,200	$ 20	$ 104,000	Lab Pack Incineration	$ 20,000	$ 84,000
Metal Precipitation	2,800	$ 20	$ 56,000	Lab Pack Treatment	$ 1,000	$ 55,000
Reactives Treatment	20	$ 100	$ 2,000	Treatment	$ 100	$ 1,900
Recycle/Reagent	1,250	$ 10	$ 12,500	Lab Pack Landfill	$ 200	$ 12,300
Separate Nonhazardous Chemicals From Hazarous	700	$ 10	$ 7,000	Lab Pack Landfill	$ ---	$ 7,000
Shock Sensitive-Detonation	77	$ 1,000	$ 77,000	Detonation	$ 100	$ 76,900
Miscellaneous Treatment	2,800	$ 20	$ 56,000	Lab Pack Treatment	$ 2,600	$ 53,400
Totals	20,627	---	$ 367,500		$ 25,350	$ 342,150

Table II. Capital Investment for Chemical Treatment

Item	Total Cost	Expected Life (YR)	Approximate 1987 Annualized Cost
Building	$ 375,000.00	30	$12,500.00
Forklift	13,500.00	15	900.00
Walk-In Hood	4,200.00	7	600.00
B/R Still	9,300.00	15	620.00
Recycleen Still	4,050.00	10	405.00
Hexane/Hemo-De Still	500.00	4	125.00
Glassware	300.00	2	150.00
Filter Press	6,600.00	20	330.00
Tank/Stand	360.00	20	18.00
Air Drive Pump	800.00	12	67.00
Electric Mixer	950.00	10	95.00
Air Drive Mixer	300.00	8	38.00
Drum Vents	800.00	3	267.00
Drum Liner	76.00	2	38.00
Process Barrel (Reactives)	42.00	2	21.00
Hand Pump (Metals)	40.00	2	20.00
Hand Pump (Solvents)	60.00	5	12.00
Transportation Drums	30.00	3	10.00
Rifle/Scope	120.00	30	$ 4.00
		Sub-Total	$16,220.00
Labor Costs for Treatment of Hazardous Wastes			$30,000.00
	Total 1987 Capital & Labor Costs		$46,220.00

department's program includes the distillation of xylene, hexane, ethyl alcohol, limonene, acetone, and lacquer thinner. For a successful distillation program, it is essential to have cooperation between the university departments. Academic researchers and laboratories must be willing to use distilled products and to be involved in keeping wastes separate. Quality control is a must. If a university is interested in setting up a distillation program, it should begin with

Table III. Operating Costs for Chemical Treatment—1987

Item	Total 1987 Cost
Acid/Base Neutralizations (Materials)	$ 600.00
Bulking Solvents	500.00
Incinerating Solvents	20,000.00
Distillation	250.00
Metal Precipitations	1,000.00
Recycle Reagents	200.00
Treat Reactives	100.00
Segregate Nonhazardous Waste	———
Detonate Shock Sensitives	100.00
Miscellaneous–Repairs, Utilities, etc.	2,600.00
Total 1987 Operating Costs	$ 25,350.00

a waste stream that is produced on a regular basis, contaminated with only one or two compounds (a single solvent or a mixture of two solvents that have substantially different boiling points).

Bulking Solvents

This operation is not really a "treatment" method, but it is a procedure universities can use to lower their disposal costs substantially. This method involves pouring laboratory size containers of spent solvent into 55 gal drums. Many solvents can be mixed together and then used as a fuel supplement in a fuels blending program, or they can be incinerated. The price for incineration of lab pack solvents is about 20 times more expensive than the price for incineration of bulk solvents. To implement this operation, one needs a ventilated area, information on the compatibility of organic solvents, 30 or 55 gal drums, and a storage area.

Incineration

The Environmental Health and Safety department upgraded an existing pathological incinerator so that it could be used to incinerate ignitable chemical waste (waste that is considered hazardous only because of its ignitability), low-level radioactive waste (solids, liquids, and animals), and pathological waste. Although this has been an effective operation for the department, it is not necessarily applicable to other universities. Because of EPA permitting requirements, it is very difficult to obtain a license for operating a new incinerator. Unless a university has an interim status permit to operate an incinerator, it may not be feasible at this time to apply for a permit for a new facility.

Metal Precipitation

This treatment method does not completely eliminate commercial disposal costs, but it does greatly reduce them. This volume reduction method involves taking aqueous heavy metal solutions and precipitating out the heavy metal in a solid sludge form. The sludge is then collected for commercial reclamation or disposal, and the aqueous solution is put down the drain. The Environmental Health and Safety department at ISU currently treats chromic acid, kjeldahl waste, photodeveloping solutions that contain silver, arsenic solutions, and mercury solutions in this way. For large volumes we use a hydraulic filter press, and for small volumes we use laboratory funnels and filter paper. We achieve about a 10 to 1 reduction in waste volume using the filter press and filter paper. The treatment methods use either sodium sulfide or sodium borohydride to form a solid metal precipitate.

Reactives Treatment

This treatment process is a relatively small operation compared to many of our other methods. But because commercial disposal companies charge a high price for the disposal of reactives, it still saves the university money. The most common reactive chemicals we treat are sodium or potassium metal. Small pieces of these metals are added to 15 gal plastic containers of water. This method causes the sodium (or potassium) to react with the water and form sodium hydroxide (or potassium hydroxide). The reaction is violent, but if done in small volumes and with the correct equipment, it is safe. The hydroxide solution is then neutralized and poured down the drain.

Recycle/Reagent

Many of the chemicals that we pick up from the university are surplus and being discarded. We sort out and save the chemicals that are in their original container and appear to be in good condition. Twice a year we publish a list of surplus chemicals that we have available for recycling. The list is distributed to departments on campus, and they call us if they are interested in any of our stock. We then deliver the chemicals to the departments at no charge.

We also use many of the surplus chemicals as reagent in our treatment methods. Examples of chemicals we routinely collect that are used as reagent are inorganic acids, sodium hypochlorite, hydrogen peroxide, sodium sulfide, potassium permanganate, and so on.

Separation of Chemicals

A wide variety of chemicals are used on a university campus. Some of the chemicals are hazardous and others are not. Chemicals that are nonhazardous can be disposed of in the normal trash, or, if they are water soluble, down the drain. Some universities handle all of the waste chemicals as hazardous waste and dispose of them at commercial hazardous waste facilities. By using separation techniques, however, funds are not wasted on the disposal of nonhazardous chemicals.

Shock Sensitive Detonation

The Environmental Health and Safety department at ISU detonates potentially explosive compounds by placing them in a 55 gal drum with a hole cut in it, and shooting the compounds with a .22 rifle. The chemicals then explode and burn. Other universities use dynamite, plastic explosive, or primer cord to detonate their explosive chemicals. Examples of explosive chemicals detonated at ISU include butyl lithium, anhydrous picric acid, ethyl ether peroxide, silane compounds, and di-nitro compounds. Extra care is required in handling, transporting, and storing of these compounds. These chemicals constitute a relatively small volume of waste compared to our other wastes, but this method saves the university a great deal of money.

Miscellaneous Treatment

There are a variety of other chemical/physical treatment methods used by the Environmental Health and Safety department at ISU to treat hazardous waste. These include oxidation of cyanide, oxidation of formaldehyde, bromine reduction, pesticide degradation, and so on. Most of these treatment methods are done in batch processes because of the small volumes.

SUMMARY

The center of the hazardous waste program at ISU is the Chemical Waste Handling Facility (CWHF). Chemical wastes generated by the university are collected by the Environmental Health and Safety department and brought to the CWHF for treatment, storage and disposal. ISU employs many different chemical/physical treatment methods to destroy, detoxify, or reduce this chemical waste. By employing these treatment methods, the university saves approximately $300,000 per year in hazardous waste disposal costs.

ACKNOWLEDGMENT

Iowa State University is indebted to the work of Emery E. Sobottka, Director of Environmental Health and Safety, who conceived the idea of building a plant where chemical/physical treatment methods could be employed to treat university-generated chemical hazardous waste, and who developed the Chemical Waste Handling Facility.

BIBLIOGRAPHY

Armour, M. A., and L. M. Browne. *Hazardous Chemicals Information and Disposal Guide* (Edmonton, Canada: University of Alberta, 1984).

________. *Potentially Carcinogenic Chemicals: Information and Disposal Guide* (Edmonton, Canada: University of Alberta, 1986).

Detoxification of Hazardous Waste (Ann Arbor, MI: Ann Arbor Science Publishers).

"Disposal of Dilute Pesticide Solutions," U.S. EPA NTIS PB-297- 985 (1979).

"Handbook for Pesticide Disposal by Common Chemical Methods," U.S. EPA NTIS PB-252–864 (1975).

"Identification and Description of Chemical Deactivation/Detoxification Methods for the Safe Disposal of Selected Pesticides," U.S. EPA NTIS PB-285–208 (1978).

National Research Council. *Prudent Practices for Disposal of Chemicals from Laboratories* (Washington, DC: National Academy Press, 1983).

Pitt, M., and E. Pitt. *Handbook of Laboratory Waste Disposal: A Practical Manual* (1985).

SECTION V

Waste Disposal Practices

CHAPTER 12

Chemical Waste Reduction and Recycling in Canadian Academic Laboratories

Linda Varangu and Robert Laughlin

INTRODUCTION

This chapter will begin by providing an introduction to the Ontario Waste Exchange (OWE), and how it got involved with helping promote waste reduction and recycling in schools. The remainder of the chapter will give examples of waste reduction and recycling programs being implemented in some Ontario schools.

The Ontario Waste Exchange program is a free help program for Ontario waste generators who would like to investigate using alternatives for disposal. The co-sponsors of the program, Ontario Waste Management Corporation (OWMC) and the Ontario Ministry of the Environment (MOE), jointly fund the OWE as part of their commitment to encourage industrial waste reduction.

The program is operated out of the Ontario Research Foundation (ORF).

INCENTIVES FOR WASTE REDUCTION

Industries are starting to find out that waste reduction and recycling may be better alternatives than waste disposal. There are three main incentives for considering alternatives to disposal: economics, legislation, and corporate image.

Current costs for disposal of both hazardous and nonhazardous wastes are increasing dramatically in southern Ontario. The small, unprepared business operation that has not had the experience of small quantity hazardous waste disposal costs may get quite a shock.

In Ontario, under the provincial legislation of the Environmental Protection Act and Regulation 309, waste generators must register their facilities and obtain manifests to transport designated waste classes to approved waste treatment/disposal facilities. This "cradle-to-grave" tracking system allows the government to obtain a better understanding of the types and quantities of wastes generated, and enables the tracking of wastes that are improperly disposed of.

Other legislation, such as the Workplace Hazardous Materials Information System (WHMIS) under the Ministry of Labour and Transportation of Dangerous Goods Act, requires the generator to obtain a greater understanding of wastes being generated.

Industries are also concerned with the corporate image aspect because bad press could also be bad for business. Smaller quantity waste generators such as schools should also be concerned because improper handling of wastes resulting in accidents could generate law suits and increases in liability insurance rates.

HOW THE OWE CAN HELP

The OWE can provide assistance in a number of different ways, some of which are summarized as follows:

1. Provide waste exchange contacts, that is, find industrial users for wastes generated in different industry sectors or help find sources of wastes as alternative raw materials.
2. Provide recycling industry contacts such as solvent, oil, and metal solution recyclers.
3. Conduct literature searches on selected waste reduction topics such as waste processing technologies.
4. Conduct mini-research projects on selected hazardous waste streams. These are funded by OWMC.
5. Provide technical assistance on waste reduction.
6. Conduct plant visits to help point out areas for improved waste management practices.
7. Offer presentations to interested groups.

WHO USES THE OWE

The primary users of the OWE are the waste generators, although a number of other associated waste management companies or organizations are also involved:

Industry:	waste generators, waste users, the service industry, chemical and equipment suppliers
Consultants:	providing a service to industries or for data collection
Government:	municipal, regional, provincial, federal agencies

THE WASTE MANAGEMENT HIERARCHY

The Ontario Waste Exchange promotes adherence to the principals in the waste management hierarchy (see Figure 1). When waste generators are evalu-

Waste Abatement	Do not make it.
Waste Minimization	If it has to be made, one should minimize its volume and toxicity.
Waste Reuse	See if someone else can use it.
Waste Recycle	If it cannot be used as is, reclaim as much as possible that is useful.
Waste Treatment	Treat what cannot be reclaimed to render it safe.
Waste Disposal	Dispose of residues to air, water or land.

Figure 1. Waste management hierarchy.

ating options to reduce waste quantities, the easiest path for them is usually treatment at the end of the pipe. However, upon careful evaluation, a more thorough understanding of their process could result in more cost effective and long-term solutions. Several examples of waste reduction approaches in industry are given in Figure 2.

CANADIAN WASTE MATERIALS EXCHANGE (CWME)

The OWE is also seen as the active component of the Canadian Waste Materials Exchange program in Ontario. The CWME is a passive waste exchange program and has been operating since 1978 with the support of federal and provincial grants and industry contributions. As a passive waste exchange, the CWME issues a bulletin every two months that contains lists of wastes available and wastes wanted by different industry sectors. Subscribers to this bulletin can browse through the booklet to see what interests them. It is, therefore, important for generators to list waste materials because no one can come up with an idea for waste utilization if he or she does not know that a waste exists. A sample listing form is provided with this paper (Figure 3).

Figure 4 shows a summary of the operation of the CWME. About 3,700 companies are participants, with close to 3,000 different wastes listed to date. Most of the wastes that are listed do receive enquiries, with about 7.4 enquiries per listing. As a passive waste exchange, the transfer rate is about 20%, which results in an annual tonnage of wastes transferred of approximately 300,000 tons. The estimated dollar value of these transferred wastes is about $11 million dollars per year.

If we analyze the information presented in Figure 5, we can see a summary of waste quantities exchanged by category. The eleventh category, after the miscellaneous column, is laboratory chemicals.

It was added as an afterthought when one industrial laboratory came to us to list the chemicals it needed to dispose of because the facility was being moved and no longer required them. We listed these laboratory chemicals—some unused, some used—in all the traditional laboratory quantities. These were transferred to a broker who found uses for them in numerous other

	Examples
WASTE ABATEMENT Substitution of a new low-waste primary industrial process for an old process to eliminate or drastically reduce the quantity of waste produced	– Replacement of sulfuric acid in steel pickling with hydrochloric acid – Replacement of liquid paints by powder coatings – Replacement of solvent based adhesives with water based adhesives
WASTE MINIMIZATION The reduction of the quantity of waste through good housekeeping practices or by the application of concentration technologies. Also the reduction in the degree of hazard of waste through simple in-plant treatment	– Separation of waste streams to permit recovery – Application of countercurrent rinsing to minimize volume discharge – Neutralization of wastes and precipitation of smaller volume sludges – Fixing leaky taps and nozzles
WASTE REUSE The direct reuse of a waste stream, as is, or with a very minor modification	– Reuse of surplus or salvage chemicals – Use of blast furnace slag as aggregate – Use of solvents from electronics industry in paints manufacture – Use of refinery spent caustic in pulping wood – Use of oil sludges in asphalt manufacturing – Use of electronic circuit manufacturing plating baths in regular plating shops
WASTE RECYCLE The reclamation of value from waste streams through the application of reprocessing technologies such as distillation, etc.	– Oil rerefining – Solvent distillation – Recovery of iron salts from pickle liquor – Recovery of heavy metals from sludges – Recovery and reuse of spent foundry sands – Recovery of scrap metal – Landfarming of organic wastes – Regeneration and reuse of activated carbon – Recycling of grease and fats to renderers

Figure 2. Waste reduction approaches.

facilities. Following this successful transfer, we became more aware of the needs of other laboratories that might also be cleaning off their shelves because of moving, general cleanup, chemicals out of date for their purpose or no longer required, etc.

canadian waste materials exchange

la bourse canadienne des dēchets

LISTING FORM

(Submit separate form for each listing)

Exchange Use Only
Code # ______
Location ______

1 Company Name ______ SIC Code: ___ ___ ___ ___

2 Mailing Address ______

3 Company Contact ______ 4 Title ______

5 Signature ______ 6 Date ______ 7 Phone ( ) ______

8 Check One Only ☐ MATERIAL AVAILABLE ☐ MATERIAL WANTED

9 Classifications (review all first, then select one only that best describes your material)

☐ ACIDS ☐ SOLVENTS ☐ PLASTICS AND RUBBER ☐ METALS AND METAL SLUDGES
☐ ALKALIS ☐ OTHER ORGANIC CHEMICALS ☐ TEXTILES AND LEATHER ☐ MISCELLANEOUS
☐ OTHER INORGANIC CHEMICALS ☐ OILS AND WAXES ☐ WOOD AND PAPER

10 Material to be listed (main usable constituent, generic name) ______

11 The industrial process that generates this waste ______

12 Main constituent (chemical formula) ______ in ______

13 Contaminants (highest to lowest, chemical formula) ___ ___ ___ ___ ___ ___ ___ % ___

14 Percentage by (check one) ☐ VOLUME ☐ WET WEIGHT ☐ DRY WEIGHT

15 Physical State ☐ SOL'N ☐ SLURRY ☐ SLUDGE ☐ CAKE
☐ AGGREGATE ☐ SOLID ☐ DUST ☐ GAS

16 Miscellaneous appropriate information (pH, toxicity, reactivity/color, particle size, flash point, total solids, **purchase date**):

17 Potential or intended use ______

18 Packaging ☐ BULK ☐ DRUMS ☐ PALLETS ☐ BALES ☐ OTHER ______

19 Present amount ______ 20 Frequency ☐ CONTINUOUS ☐ VARIABLE ☐ ONE TIME

21 Quantity thereafter ______ in ☐ lbs ☐ tons ☐ gal ☐ C y ☐ Kg. ☐ Tonnes
☐ Litres ☐ Cubic Meters ☐ Other ______

per ☐ Day ☐ Week ☐ Month ☐ Quarter ☐ Year

22 Restrictions on amounts ☐ none ☐ minimum ☐ maximum ______

23 Available to Interested Parties ☐ sample ☐ lab analysis ☐ independent analysis

24 For material wanted, acceptible geographic range (i.e., states, provinces, regions, countries)

25 For material available, if location of the material differs from the above mailing address, complete the following:

City ______ State or Province ______ Area Code ______

26 If necessary to speed communication, please check if your company name, address and telephone number may be released ☐ YES ☐ NO

2888H 27

2395 Speakman Drive, Sheridan Park Research Community, Mississauga, Ontario, Canada. L5K 1B3 (416) 822-4111

Figure 3. Sample listing form.

We started promoting the waste exchange concept to other laboratories, in industries, and in schools. We completed a video with the National Film Board that highlights waste exchange and waste reduction, and will be distributed to high schools across Canada. Various school systems now include waste reduction as part of their curriculum, and a package designed for elementary school children can be obtained from the MOE in the near future that includes a waste exchange game for students to learn with.

HAZARDOUS WASTE GENERATION IN ONTARIO SCHOOLS

In Ontario, there are a total of 700 schools that have registered as hazardous waste generators with the MOE. Table I shows different waste types that have been registered and the quantities of tonnes disposed of. As can be seen, out of the 1,818 tonnes of hazardous wastes registered by schools, 1,200 tonnes of

January 1, 1978, to March 31, 1988	
Number of Participating Companies	3,500
Number of Wastes Listed	2,887
Number of Wastes Enquired About	2,574 (89.2% of listings)
Number of Enquiries	20,842 (7.2 per listing)
Number of Wastes Transferred	539 (18.7% of listings)
Annual Tonnage of Wastes Transferred	320,500 tons
Value of Wastes Transferred	$11.8 million/year

Figure 4. Summary of the operation of the Canadian Waste Materials Exchange.

waste oils and lubricants, oil skimmings and sludges, and halogenated pesticides are generated, leaving ~600 tonnes of other wastes, of which only 97 tonnes are organic and inorganic laboratory chemicals. Considering that Ontario produces about 3.6 million tonnes of hazardous wastes per year, this number seems minuscule. However, the generator, who was not paying high costs for disposal of laboratory wastes in the past, is now facing these costs, which must be absorbed somehow into the system.

ALTERNATIVES FOR WASTE CONTROL IN SCHOOL LABORATORIES

We at the Canadian and Ontario exchanges have definite reservations as to whether a passive information exchange can effectively handle laboratory chemicals or whether these are better handled in other ways, such as regional collection and redistribution centers or through active chemical brokerages.

Internal Waste Exchange System

As with all waste-reuse approaches, the first place to look for possible users for surplus laboratory chemicals is within the company or institute that first

To March 31, 1988

Category	Tons/Year
1. Organic Chemicals	2,229.7
2. Oils, Fats & Waxes	125,432.5
3. Acids	11,886.1
4. Alkalis	13,385.1
5. Other Inorganic Chemicals	3,297.0
6. Metals & Metal Sludges	16,245.0
7. Plastics	1,788.2
8. Textiles, Leather & Rubber	5,775.1
9. Wood & Paper Products	82,517.0
10. Miscellaneous	57,901.0
11. Laboratory Chemicals	0.1
	320,456.8

Figure 5. Summary of waste quantities exchanged by category.

Table I. Hazardous Wastes Generated in Ontario Schools

MOE Waste Code	Description	# Schools Registered	Total Quantities (tonnes)
112	Acid Wastes, Heavy Metals	44	11
113	Acid Wastes, Other Metals	3	7
114	Other Inorganic Acids	3	10
121	Alkaline Wastes, Heavy Metals	17	1
122	Alkaline Wastes, Other Metals	8	45
131	Neutralized Wastes, Heavy Metals	1	–
145	Paint/Pigment/Coating/Residues	4	7
148	Inorganic Laboratory Chemicals	472	62
211	Aromatic Solvents	7	17
212	Aliphatic Solvents	8	46
213	Petroleum Distillates	68	59
221	Light Fuels	6	24
232	Polymeric Reins	1	85
241	Halogenated Solvents	3	36
242	Halogenated Pesticides	1	209
251	Oil Skimmings & Sludges	7	366
252	Waste Oils & Lubricants	33	708
253	Emulsified Oils	4	24
263	Organic Laboratory Chemicals	1	35
264	Photo Processing Wastes	7	63
265	Graphic Art Wastes	1	2
	Totals	700	1,818

produced the excess chemical. In a recent study (1979) by Mrs. M. C. Holton of the River Road Laboratories of Environment Canada, "Recommended Practices for the Management of Hazardous Wastes at Institutions," the following was the recommended practice for the operation of an internal waste exchange:

> Waste Exchange
> Chemical wastes generated by institutions are very often in a usable form, many times in an unopened state. The chemicals have been discarded either because too large an amount had been ordered, there had been changes in projects or personnel, or in a general cleaning out of inactive chemicals. Larger institutions should consider establishment of an internal waste exchange, to make use of existing assets, to reduce disposal trouble and expense, and to retain stocks of rare, possibly irreplaceable chemicals.
>
> The decision to establish and operate an internal waste exchange need not be an "all or nothing" approach. Depending on management policies, available facilities, and work areas, a larger research establishment may opt to retain the majority of potentially useful chemicals that can be held safely and economically. A school board or smaller institution may decide to retain only the least hazardous chemicals and only demonstration quantities of some other chemicals. It is poor practice to discard a chemical being held safely when a future need is known, especially if a larger quantity of chemical may have to be acquired for replacement.
>
> Whatever the degree of waste retention, only stable chemicals should be kept in

the waste exchange storage area. One person should be responsible for accepting chemicals, maintaining an inventory, and supervising the area. This waste exchange storekeeper should report to the hazardous waste manager and have sufficient training and experience to recognize the majority of chemical types being retained, and be able to use handbooks on chemical and reactive hazards. In many institutions, this role can be assumed by an existing storekeeper.

Waste exchange materials should be held in a properly serviced area, separate from other chemical storage areas. The waste exchange area should be well ventilated and designed to minimize fire and explosion hazards by following recommendations of the National Fire Protection Association and other authorities. Reactive, incompatible chemicals should be segregated as in storerooms.

Each institution operating a waste exchange must set procedures for transferring chemicals into and out of the storage area. The following procedures are offered as a guideline:

(1) Materials to be stored must be transferred by the donating group to the waste exchange storekeeper. The chemical must be labelled following an approved procedure. In addition, information must be given on the suspected purity of the chemical and possible contaminants, its age, whether the material has been opened and the time since being opened. The storekeeper has the authority to refuse a material for improper labelling or for any other reason. If the storekeeper is unfamiliar with a chemical, he should refer to handbooks on chemical and reactive hazards, or consult with more knowledgeable persons before accepting the chemical.

(2) The chemical is assigned an inventory number and this is marked on the label. The chemical is then placed at an appropriate spot in the storage area.

(3) The information from the label and the inventory number are entered in the waste exchange inventory listing, maintained either on a card system or preferably, on a computer system. The purchasing department should have copies of the card system or access to the computer inventory. Purchasing agents should encourage requisitioners to consider use of the stored chemicals rather than making new purchases.

(4) The inventory should be reviewed periodically by the hazardous waste manager to ensure that adequate records are kept, and that only appropriate chemicals are accepted. The storekeeper is responsible for keeping the inventory up-to-date.

(5) The storekeeper should examine the deposited chemicals at least every 6 months to check for degradation or leakage. Any decomposing chemicals should be disposed of immediately. Prior to the disposal of a leaking container, the storekeeper should contact the donating group. If the donating group decides the chemical should be retained, it should arrange for transfer to a proper container. Otherwise, the storekeeper should arrange for disposal.

(6) Staff wishing a chemical from the storage area must make a request to the waste exchange storekeeper, who, using established criteria, may or may not issue the material. The requisitioning group is responsible for pick-up and transport.

(7) The entry in the inventory for a chemical removed from the storage area can be handled in two ways. The storekeeper may remove the card or computer entry for the chemical, noting to whom and when the transfer was made, and holding the card for reporting to the hazardous waste manager. Alternately, the card or computer entry can be altered to show the transfer, but still retained. Using this procedure, an individual searching for a chemical within the institution could be referred to the recipient group.

A waste exchange operated with the participation of several institutions in a district could be workable in many cases. For example, several research institutes located in a research park would generate a sufficient volume of potentially useful chemicals to warrant a shared waste exchange facility. Such a common facility could be operated by one of the institutes, with costs distributed to all users.

This concept of a regional laboratory chemicals waste exchange facility has been explored by the federal government of Canada for the national capital area to serve federal laboratories.

Regional Collection and Chemical Brokerages

For laboratory chemicals, it seems to be important to have a central collection point from which they can be redistributed, and which offers potential purchasers some assurances on quality of the materials. For major industrial waste streams available on an on-going basis, a company is willing to invest some time and effort to ensure themselves that the material is suitable for their application. For jars of laboratory chemicals, this is not feasible, so that confidence must be developed in the potential purchaser that the operators of the laboratory chemical exchange will give assurances of the quality of the chemicals being supplied.

Zero Waste Systems, Inc.

Zero Waste Systems, Inc. (ZWS) of California has been in the business of recycling and disposing of laboratory chemicals for a number of years. In 1982, it sold about 100,000 lb of laboratory chemicals. The following notes on its operation are based on personal communications with Trevor Pitts, past president of Zero Waste Systems, and Robert Laughlin of the Canadian Waste Materials Exchange:

- Approximately half the containers of chemicals collected are recycled.
- Solvent wastes are bulked and recycled.
- Bottles of chemicals are classified into "dump" and "inventory" classes. The "dump" criteria are:
  a. bad, partially missing, or handwritten labels
  b. cracked or unsuitable caps
  c. containing less than half the labeled quantity

   d. obvious quality problems, e.g. water in deliquescent or reactive materials
   e. caked reacted materials
   f. excess already in inventory
- Many chemicals are fine for long term storage, and those that are not are usually obvious because the deterioration is visible, or because the chemistry is well known (e.g., peroxides in ethers).
- To be successful, the company has to be a "retail laboratory chemical operation." That means new chemicals are needed to keep up inventory where there are no wastes, glassware, or equipment. An entire business in recycled material will probably only average half the sales because the most commonly used chemicals are less often disposed. ZWS often gets used glassware along with chemicals for disposal.
- The geographic range that can be best served by such a "retail laboratory chemical operation" is about a 100-mile radius. Some bulk sales are made to wholesalers beyond that range, and ZWS is looking into exporting to Third World school systems.
- Passive waste exchange does not work in laboratory chemical reuse because of the quality assurance problems.
- Catalogs do not work. By the time they are distributed, they are out of date. In addition, they are expensive to produce because of the extensive listings of small quantities of materials required.

These comments are from a highly entrepreneurial type of operation, but they are based on nearly a decade of real experience in this area. The company is not too positive about the possibilities of the passive waste exchange playing some role in laboratory chemicals reuse. However, use of personal computers can speed up information transfer.

For laboratory chemicals, it would seem that local approaches to their reuse and recycling would have the greatest chance of succeeding: in-house exchanges for larger institutions and community exchanges for groups of laboratories that are geographically close to one another. For other laboratories, perhaps the waste disposal companies now specializing in laboratory chemicals could be encouraged to undertake more recycling activity along the lines described by Zero Waste Systems. Another alternative would be to query the existing major laboratory supply houses to see if they would consider taking back chemicals and reselling them as second-hand products.

EXAMPLES OF WASTE REDUCTION AND RECYCLING IN SCHOOLS

Example 1—Case Study

The Niagara South School Board in southern Ontario comprises 15 high schools, 60 elementary, 2 maintenance buildings, and a central storage building. It covers an area of 400 square miles. Mr. Stewart Franck was the safety officer employed by the Board and he had also shown a personal commitment to waste reduction and recycling.

Step 1	Waste Audit By: Science teachers and Safety Officer
Step 2	Set Up Centralized Disposal and Inventory Control By: Safety Officer
Step 3	Disposal Alternatives By: Safety Officer

Figure 6. Waste management program.

Mr. Franck began an inventory and disposal program by having waste audits conducted in all the schools. The people responsible for these audits in the schools were primarily the science teachers. A survey by the safety officer followed (Figure 6).

Mr. Franck then set up a centralized disposal and inventory control system. The centralization of these two activities appears to be the key to minimizing costs, according to Mr. Franck. The school waste disposal coordinators advise Mr. Franck of the materials they have available for disposal. Mr. Franck, who is aware of the chemical needs in other schools within the Board, evaluates the material destined for disposal and transfers material he considers useful to qualified persons at other schools. Another key point is the establishment of lines of communication between the teachers receiving the so-called wastes, the collector, and the generator.

The material destined for disposal is stored until the other schools have significant quantities as well, at which time a small quantity waste disposal company picks up from all the schools in one run. Mr. Franck estimates savings are in the $10,000 range.

A summary of hazardous waste materials collected from the Niagara South School Board is given in Table II, along with their disposal alternatives.

Example 2—Reuse of Industrial Chemicals in School Laboratories

School laboratories can also reduce costs by acquiring some of their laboratory chemical needs from local industries—either from the laboratories or other manufacturing processes. In Ontario, several university laboratories are involved in this type of exchange. Large pharmaceutical industrial laboratories who have very strict quality control/quality assurance requirements on their chemicals are required to dispose of these chemicals regularly. Storage life for such chemicals is actually much longer for general use, and some university laboratories are quite grateful to receive these materials—usually at no cost. The most important requirements in this type of an exchange are the establishment of trust between the recipient and generator and the constant good quality of materials.

Table II. Waste Materials Collected – 1987

Material	Quantity/ Year	Disposal Alternatives
Antifreeze	960 L	
Car Batteries	50	Looking for recycler to pick up
Gas Cylinders	50	Sent back to manufacturer for credit
Inorganic Laboratory Wastes (misc.)	60–240 L	Several hundred bottles of chemicals, acids, solvents reused by other schools in Board system
Organic Laboratory Wastes (misc.)	~240 L	
Oils (crankcase)	3–4000 L	Off-site reclamation
Oils (vegetable)	N/A	Off-site recycling to renderer
Solvents	N/A	Exploring solvent substitutes, on-site distillation units, off-site recovery
Pesticides/Herbicides	240 L	Tried to find reusers (horticulture schools, etc.) but material too old

Note: Estimated savings by recycling: $10,000.

Example 3—Reuse of School Laboratory Chemicals in Industry

When laboratory chemicals from schools are of reusable quality, but cannot be used within the school or school board system, other options could be explored:

- Suppliers — Contact the original supplier of unused chemicals to see if they can be returned for credit or exchange.
- Brokers — Contact chemical brokers in the area to see if they might be interested in taking a look at some chemicals that may be reusable. Brokers may be interested because they often know of other organizations that might require the chemicals.
- Small Quantity Users — Find out in what type of manufacturing the material could be used by consulting the *Merck Index*. Contact various small quantity users in the area. These could be local artists, small manufacturers, or research facilities. As an example, a group of artisans using metal salts for ceramic glazes requires small quantities.
- Source Industries — Find out if there is a pickup and disposal company in the area that combines this service with recycling. Reusable chemicals could be separated from the waste chemicals and resold. Perhaps such a service could be started in your area.
- Waste Exchanges — Waste exchanges are located all over North America

and may be able to give some pointers on finding market alternatives for reusable chemical wastes. See Appendix I for a list of nonprofit waste exchanges in North America.

CONCLUSION

Waste exchange and waste reduction can occur in school laboratories as part of the overall waste management planning and audit process, and the keys to success in this area appear to be:

1. putting someone in charge of the program;
2. centralizing disposal and purchasing operations;
3. recognizing that all wastes are not created equal.

APPENDIX I

Waste Exchanges Operating in North America (April 1989)

Alabama Waste Exchange
Mr. William J. Herz
The University of Alabama
Post Office Box 870203
Tuscaloosa, AL 35487-0203
(205) 348-5889
FAX: (205) 348-8573

Alberta Waste Materials Exchange
Mr. William C. Kay
Alberta Research Council
Post Office Box 8330
Postal Station F
Edmonton, Alberta
CANADA T6H 5X2
(403) 450-5408

California Waste Exchange
Mr. Robert McCormick
Department of Health Services
Toxic Substances Control Division
Alternative Technology Section
Post Office Box 942732
Sacramento, CA 94234-7320
(916) 324-1807

Canadian Chemical Exchange*
Mr. Philippe LaRoche
P.O. Box 1135
Ste-Adele, Quebec
CANADA J0R 1L0
(514) 229-6511

Canadian Waste Materials Exchange
ORTECH International
Dr. Robert Laughlin
2395 Speakman Drive
Mississauga, Ontario
CANADA L5K 1B3
(416) 922-4111 (Ext. 265)

Enstar Corporation *
Mr. J. T. Engster
P.O. Box 189
Latham, NY 12110
(518) 785-0470

Georgia Waste Exchange *
Mr. Michael Wheelus
c/o American Resource Recovery
Marietta, GA 30050
(404) 363-3022

Great Lakes Waste Exchange
Ms. Kay Ostrowski
400 Ann Street, N.W.
Suite 201-A
Grand Rapids, MI 49504-2054
(616) 363-3262

Indiana Waste Exchange
Ms. Susan Scrogham
P.O. Box 1220
Indianapolis, IN 46206
(317) 634-2142

Industrial Materials Exchange Service
Ms. Diane Shockey
Post Office Box 19276
Springfield, IL 62794-9276
(217) 782-0450
FAX: (217) 524-4193

Industrial Waste Information Exchange
Mr. William E. Payne
New Jersey Chamber of Commerce
5 Commerce Street
Newark, NJ 07102
(201) 623-7070

Manitoba Waste Exchange
Mr. James Ferguson
c/o Biomass Energy Institute, Inc.
1329 Niakwa Road
Winnipeg, Manitoba
CANADA R2J 3T4
(204) 257-3891

Montana Industrial Waste Exchange
Mr. Don Ingles
Montana Chamber of Commerce
P.O. Box 1730
Helena, MT 59624
(406) 442-2405

New Hampshire Water Exchange
Mr. Gary J. Olson
c/o NHRRA
P.O. Box 721
Concord, NH 03301
(603) 224-6996

Northeast Industrial Waste Exchange
Mr. Lewis Cutler
90 Presidential Plaza, Suite 122
Syracuse, NY 13202
(315) 422-6572
FAX: (315) 422-9051

Ontario Waste Exchange
ORTECH International
Ms. Linda Varangu
2395 Speakman Drive
Mississauga, Ontario
CANADA L5K 1B3
(416) 822-4111 (Ext. 512)

Peel Regional Waste Exchange
Mr. Glen Milbury
Regional Municipality of Peel
10 Peel Center Drive
Brampton, Ontario
CANADA L6T 4B9
(416) 791-9400

RENEW
Ms. Cheryl Wilson
Texas Water Commission
Post Office Box 13087
Austin, TX 78711-3087
(512) 463-7773
FAX: (512) 463-8317

Southeast Waste Exchange
Ms. Mary McDaniel
Urban Institute
UNCC Station
Charlotte, NC 28223
(704) 547-2307

Southern Waste Information Exchange
Mr. Eugene B. Jones
Post Office Box 960
Tallahassee, FL 32302
(800) 441-SWIX (7949)
(904) 644-5516
FAX: (904) 574-6704

Wastelink, Division of Tencon Inc.
Ms. Mary E. Malotke
140 Wooster Pike
Milford, OH 45150
(513) 248-0012
FAX: (513) 248-1094

* For-Profit Waste Information Exchange

CHAPTER 13

Lab Pack Management

Joyce A. Kilby, Daniel L. Holcomb, and Jennifer M. Kinsler

INTRODUCTION

There was a day when chemists and laboratory technicians disposed of their laboratory waste by simply pouring it down the drain or throwing it into the trash. Even though laboratories account for less than 1% of all hazardous waste generated, they produce a wide variety of different wastes in much smaller quantities (*RCRA and Laboratories,* p. 5). This has made the temptation even greater to dispose of waste down the drain. But those practices and days have gone. They have been replaced by a practical, efficient, but above all, responsible means of laboratory waste disposal—the lab pack. Approximately 100,000 lab packs are used each year in the U.S. with a cost to the user of between $250 to $350 per pack to dispose (*Federal Register*, 51, p. 46826)

Just what is a lab pack? It is a recognized packaging unit of the U.S. Department of Transportation (DOT) that allows different materials from the same hazard class to be packaged together in specified containers for treatment and disposal. Historically, the lab pack was the concept of GSX Chemical Services, Inc. In 1978 they requested and received an exemption (DOT E129) allowing corrosive liquids in small containers to be surrounded by an absorbent (*Chemical Engineering Progress*, p. 10) and packed in steel drums. This exemption has since been converted into a general rule and appears in the *Code of Federal Regulations*, 49 CFR 173.12, "Exceptions for Shipment of Waste Material."

Another governmental authority to be recognized in lab pack management is the U.S. Environmental Protection Agency (EPA), acting specifically under the guidelines of the Resource Conservation and Recovery Act (RCRA) of 1976. This act was designed to protect human health and the environment from improper management of solid and hazardous waste through a manifest and record-keeping system. This system tracks the movements of hazardous waste from generation to final disposal. In May 1980, the EPA promulgated a set of RCRA regulations covering large quantity generators (1000 kg or more per calendar month) and in September 1986 promulgated a new set of RCRA regulations for small quantity generators (100–1000 kg/month) previously exempt. This regulation moved many of the small laboratories into the hazardous waste regulatory spotlight. It is estimated that there are over ten times more small quantity generators than large quantity generators, and most of them are as yet unaware of the need for compliance (Coffey, p. 23).

What does all of this mean in the pursuit of lab pack management? It means the laboratory must understand and comply with the applicable elements of EPA RCRA and DOT regulations. While wastes generated by laboratories (spent solvents, unused chemicals, reaction products, and testing samples) may not appear on any of the EPA regulated waste lists, they may be regulated because of their hazardous characteristics (ignitability, corrosivity, reactivity, and EP toxicity).

It should be noted that the material presented in this chapter is based on federal DOT and EPA regulations. These federal regulations do not prohibit states from adopting more stringent hazardous waste regulations. To determine responsibility, it is necessary to check with state and local authorities. Table I lists the telephone numbers of state solid and hazardous waste agencies ("RCRA and Laboratories," p. 23).

PRE-SHIPMENT ACTIVITIES

The first step in pre-shipment is to classify the laboratory chemical wastes into compatible groups. The DOT (49 CFR 173.12) labels these groups flammable liquid, flammable solid, oxidizer, corrosive, Poison B, Otherwise Regulated Material-A (ORM-A), ORM-B, ORM-C and ORM-E. Other DOT requirements imposed include:

- Wastes from only one DOT hazard class may be packed in one shipping container.
- Wastes packaged in the lab pack must be compatible with each other, compatible with the packaging used, and with the container itself.
- Inside packaging of liquid must be surrounded by a compatible material capable of absorbing the total liquid contents. This material must also serve as a shock absorber.

After determining the hazard class, the generic shipping name from the Hazardous Materials Table (49 CFR 172.101) may be used in place of specific chemical names when two or more chemically compatible wastes from the same hazard class are packed.

EPA restrictions should be examined during the pre-shipment phase. Laboratory wastes identified in any of the EPA restricted lists means the lab pack may not be land disposed unless the restricted wastes are removed and treated according to specific standards found in the *Code of Federal Regulations* 40 CFR 268 Subpart D or a successful petition has been demonstrated under 40 CFR 268.6. These EPA restricted lists include the California List, F001-F005 Solvent Ban and Dioxin Waste Ban, First Third, Second Third and Third Third lists.

Table I. State Solid and Hazardous Waste Agencies: A listing of telephone numbers of state agencies responsible for hazardous waste management and disposal

State	Telephone	State	Telephone
Alabama	(205) 271–7737	Montana[a]	(406) 444–2821
Alaska	(907) 465–2666	Nebraska[a]	(402) 471–4217
Arizona[a]	(602) 257–2211	Nevada[a]	(702) 885–4670
Arkansas[a]	(501) 562–7444	New Hampshire[a]	(603) 271–4608
California	(916) 322–2337	New Jersey[a]	(609) 292–8341
Colorado[a]	(303) 320–8333	New Mexico[a]	(505) 827–0020
Connecticut	(203) 566–5712	New York	(518) 457–3274
Delaware[a]	(302) 736–4781	North Carolina[a]	(919) 733–2178
Florida[a]	(904) 488–0300	North Dakota[a]	(701) 224–2366
Georgia[a]	(404) 656–2833	Ohio	(614) 466–7220
Hawaii	(808) 548–6767	Oklahoma[a]	(405) 271–7266
Idaho	(208) 334–5879	Oregon[a]	(503) 229–5913
Illinois[a]	(217) 782–6760	Pennsylvania	(717) 787–7381
Indiana[a]	(317) 232–4458	Rhode Island[a]	(401) 277–2797
Iowa	(515) 281–8308	South Carolina[a]	(803) 758–5681
Kansas[a]	(913) 862–9360	South Dakota[a]	(605) 773–3329
Kentucky[a]	(502) 564–6716	Tennessee[a]	(615) 741–3424
Louisiana[a]	(504) 342–1227	Texas[a]	(512) 463–7830
Maine	(207) 289–2651	Utah[a]	(801) 538–6170
Maryland[a]	(301) 225–5709	Vermont[a]	(802) 828–3395
Massachusetts[a]	(617) 292–5500	Virginia	(804) 225–2667
Michigan	(517) 373–2730	Washington	(206) 459–6301
Minnesota[a]	(612) 296–7373	West Virginia	(304) 248–5935
Mississippi[a]	(601) 961–5062	Wisconsin[a]	(608) 266–0833
Missouri[a]	(314) 751–3176	Wyoming	(307) 777–7752

[a]States having EPA authorized state hazardous waste programs.

The California List is based on regulations developed by the California Department of Health Services. Effective 8 July 1987, it prohibits the disposal by land of the following wastes listed under RCRA Section 3001:

1. liquid hazardous wastes associated with solids or sludge containing free cyanides at concentrations greater than or equal to 1000 mg/L;
2. free liquids including any solid or sludge that contains the following metals or compounds of these metals at concentrations greater than or equal to what is specified below:
   arsenic (500 mg/L);
   cadmium (100 mg/L);
   chromium VI (500 mg/L);
   lead (500 mg/ L);
   mercury (20 mg/L);
   nickel (134 mg/L);
   selenium (100 mg/L);
   thallium (130 mg/L);
3. liquid hazardous wastes that have a pH of less than or equal to 2.0;
4. liquid hazardous wastes containing polychlorinated biphenyls (PCBs) at concentrations greater than or equal to 50 ppm;
5. hazardous waste that contains halogenated organic compounds in total concentrations greater than or equal to 1000 mg/kg.

As of 8 August 1988, under RCRA, the disposal of a California List Waste is also prohibited by deep well injection.

Hazardous wastes containing solvent and numbered F001-F005 in 40 CFR 261.31 are banned from land disposal as of 8 November 1986. Also banned on that date are the dioxin-containing wastes numbered F020-F023 in 40 CFR 261.31. Disposal of these wastes is prohibited by deep injection wells as of 8 August 1988.

The last three restricted lists, First Third, Second Third, and Third Third, are a result of the 1984 Hazardous and Solid Waste Amendments (HSWA) added to RCRA. Under this law, EPA is directed to issue standards for chemical and/or physical treatment of wastes (Pettit, p. 30). These standards would have to be met prior to land disposal. Deadlines for the standards are First Third, 8 August 1988; Second Third, 8 June 1989, and Third Third, 8 May 1990.

The First Third wastes that should have been evaluated by 8 August 1988 are listed in 40 CFR 268.10. Treatment standards were set for wastes that represented the greatest volumes and/or the greatest hazard. Usually, treatment standards are performance-based. Any technology can be used in waste treatment as long as it is the Best Demonstrated Available Technology (BDAT).

First Third and Second Third wastes that may not have treatment standards promulgated by their deadlines will be subject to certain Soft Hammer restrictions under the law (Pettit, p. 32). The Soft Hammer wastes can be disposed of in a landfill or surface impoundment only if the facility meets minimum technology requirements (MTR), and if certification exists that treatment is unavailable (Pettit, p. 32). On 8 May 1990, the Hard Hammer will go into effect. This means listed wastes will be banned from land disposal unless the facility has demonstrated that no migration of the wastes into the environment will take place.

All of the EPA restrictions must be carefully evaluated under the law. The land disposal restrictions are complicated but must be considered before lab packing wastes. Table II lists Hotline telephone numbers for assistance.

At this point in lab pack management, chemical compatibilities should be addressed. Chemical waste incompatibility can result in one or more of the following: heat generation, fire, explosion, toxic gas or fume formation, volatilization of toxic or flammable substances, formation of substances of greater toxicity, formation of shock or friction sensitive compounds, pressure, dispersion of toxic dusts, mists and/or particles, and violent polymerization. As this list indicates, there is a need to identify and segregate such incompatible wastes as unknowns, peroxide formers, and shock sensitives.

Unknowns are of particular concern because the hazard class of the waste must be known prior to lab packing. Chemical analysis is therefore necessary to determine the identity of the chemical waste. There are private laboratories that will analyze unknowns for a fee, as well as companies that produce qualitative test kits to assist in the identification of an unknown waste.

Compounds that form peroxides may be explosive. Those that form explo-

Table II. Hotlines

EPA RCRA Hotline
Provides information regarding RCRA and Superfund
1-800-424-9346

DOT Hotline
Provides information on federal regulations pertaining to the transportation of hazardous materials.
1-202-366-2993

EPA Small Business Hotline
Provides information for small quantity generators of hazardous waste.
1-800-368-5888

American Society for Testing Materials (ASTM)
1-215-299-5400

Occupational Safety and Health Administration
1-202-523-7075

sive peroxides in storage include divinylacetylene, diisopropyl ether, and sodium amide. These chemicals should be tested for peroxide content quarterly, and positive samples should be disposed of safely. Liquids in which a degree of concentration is necessary before hazardous levels of peroxide will develop include common solvents containing one or two ether functions, such as diethyl ether tetrahydrofuran, dimethyl ethers of ethylene glycol or diethylene glycol, or susceptible hydrocarbons such as propyne, cyclohexene and butadiene. These should be tested at yearly intervals and repurified or disposed of as necessary. Especially dangerous are peroxidizable monomers where the presence of peroxides may initiate exothermic polymerization, such as acrylic acid, styrene, methyl methacrylates, and acrylonitrile. In general, all peroxidizable compounds should be tested prior to lab packing to determine whether dangerous levels of peroxide exist in their individual containers.

Shock sensitive materials are those that may detonate when exposed to heat, friction, or impact. Some of these materials, picric acid or trinitrophenol, for example, are considered high explosives in their pure form and should be handled as such with extreme care taken to avoid any conditions that may lead to detonation. The manufacturer of the material should be contacted prior to disposal to determine if lab packing is appropriate.

The EPA Chemical Compatibility Guidance Document, prepared under the jurisdiction of the American Society for Testing and Materials (ASTM) Committee D-34, was developed based on case histories of accidents resulting from insufficient or inaccurate information concerning waste characteristics, handling, and disposal practices. The method of determining waste compatibilities has as its principal assumption that waste interactions are caused by reactions between the pure chemicals in the wastes. Included in this assumption is the condition that the chemicals react at ambient temperature and pressure, and that their reactivities are unaffected by concentration, synergistic, and antagonistic effects. The pure chemicals known or expected to be present in the hazardous wastes are classified under 41 reactivity group numbers based on molecular functional groups or chemical reactivities.

The guide provides a systematic method for determining binary incompatibilities for hazardous wastes, classifies the compounds and wastes into reactivity groupings, and lists potential reactions of most incompatible binary combinations of the groupings. It consists of a step-by-step procedure and an incompatibility chart—the chart being the key element. The guide cautions, however, that the incompatibility chart can only be used with the guide.

The EPA Chemical Compatibility Guidance Document is applicable to four waste categories: known compositions, compositions whose chemical class or reactivities are known, compositions known by generic name, and unknown compositions requiring chemical analysis. It also includes a flow chart, lists, worksheet, and the incompatibility wall chart. It is available from the ASTM (see Table II).

Once the segregation, identification, and classification of the laboratory wastes has been completed, creating a lab pack inventory sheet is an effective method of keeping track of the waste chemicals. A separate list can be maintained for each hazard class. The inventory sheet should contain chemical name, hazard class, size/quantity of containers, number of containers, United Nations/North American (UN/NA) classification number, EPA Waste Type Number, Reportable Quantities (RQ), and total amount of chemicals. A copy of this inventory should be included on top of the final absorbent layer of the lab pack; additional copies should be provided for the disposal facility and kept for the generator's records (Newton, p. 24). The lab pack inventory sheet will provide a means to effectively and efficiently handle and package lab wastes in a cost effective manner.

PACKAGING AND PREPARING FOR SHIPMENT

When pre-shipment identification, analysis, and segregation has been made, the actual packaging and preparation for shipment can proceed. Several types of outside packagings are used for lab packs: DOT specified metal or fiber drums. Polyethylene drums may also be used if they meet specific vibration, compression, and drop tests (49 CFR 173.12). Inside packaging must be either glass not exceeding 1 gal, or metal or plastic packagings not exceeding a rated capacity of 5 gal. In addition to packaging requirements listed above, gross weight may not exceed 450 lb or the rated capacity of the drum, whichever is less.

Prohibited materials for lab packing include acrolein, bromine pentafluoride, bromine trifluoride, chloric acid, chlorine trifluoride, fuming nitric acid, pyroforic liquids, and fuming sulfuric acid.

Tips for filling the lab pack include putting only one hazard class in each lab pack; putting larger labeled containers on the bottom, smaller toward the top;

keeping containers in upright position; and placing a compatible absorbent around containers. A packing list, to be included with the shipment and to assist in filling out the Hazardous Waste Manifest, should be prepared as the containers are placed in the drum. A final layer of absorbent should be added to top off the drum.

Waste materials for transport off-site must meet the following DOT requirements:

- They must be packaged according to DOT regulations. For lab packs, 49 CFR 173.12.
- They must be labeled according to DOT regulations under 49 CFR Part 172 and marked according to 49 CFR 172.304. Lab packs need a "This End Up" label also.
- They must be placarded on shipping vehicles according to 49 CFR Part 172, Subpart F.

The Uniform Hazardous Waste Manifest as well as the hazardous waste label are required to be completed and to accompany the waste shipment. The Manifest contains, among other data, the DOT description of the waste (hazard class), the quantity of waste, and the number of containers. Additional information on hazardous waste manifesting and labeling is found in the EPA *Code of Federal Regulations* 40 CFR 262.

TREATMENT AND DISPOSAL ALTERNATIVES

Most laboratories will ship their wastes off-site to an EPA approved landfill or treatment facility ("RCRA and Laboratories," p. 18). Many of the treatment, storage, or disposal (TSD) facilities have special requirements for wastes that must be met prior to shipment. In selecting a facility, it is not only necessary to know what type of wastes that facility will accept, but also to ascertain that the company is reliable and responsible. The object is to choose a treatment, storage, and disposal facility as well as a hauler to assure proper and legal handling of wastes. Check the reputation of the firm, how many years it has been in business, and whether it is involved in any legal or court actions. Trade associations and the local Chamber of Commerce are a valuable resource.

Until several years ago, secure chemical landfills were the chosen alternative for waste disposal. While landfilling is still currently the least expensive option in most areas, long-term liability, limited space, and current environmental regulations require facilities to examine alternative disposal methods ("RCRA and Laboratories," p. 18). Cost is a factor to be evaluated, also. For lab pack disposal in less than truck loads, the costs can range from $300 to $600 per packed 55-gal drum ("RCRA and Laboratories," p. 21). Incineration, on the average, is the most expensive ("RCRA and Laboratories," p. 21). Industry spends an estimated $5 billion treating hazardous waste annually with that amount expected to double by 1990 (Cheremisinoff, p. 3). By 1990, an estimated 280 million metric tons of waste will be treated in the U.S. (Cheremisinoff, p. 3).

There are three treatment and disposal technologies to be addressed: incineration, chemical treatment, and solidification through encapsulation. Incineration is the thermal treatment of hazardous waste using high temperatures to change chemical, physical, or biological character or composition ("RCRA and Laboratories," p. 20). Typical combustion temperatures range from 800° to 3000° F (Cheremisinoff, p. 119). For the laboratory and lab packs, incineration can be used to destroy organic hazardous wastes safely ("RCRA and Laboratories," p. 20). Incineration, however, is on the average the most expensive treatment and disposal technology. Multiple hearth furnaces, rotary kilns, or liquid injection systems are typical incineration systems used today.

In chemical treatment processes, hazardous wastes are altered by chemical reactions that may destroy the hazardous component or render the still hazardous material more convenient for further processing. Problems in using these processes are due to the limited solubilities of some metals, reaction-inhibiting impurities, and the potential for generating hazardous by-products. Chemical treatment may include solidification, neutralization, oxidation, reduction, hydrolysis, and precipitation. Chemical treatment of laboratory wastes may reduce the use of landfilling and limit potential liability. However, off-site chemical treatment can be more expensive because of the variety of wastes and amount of handling ("RCRA and Laboratories," p. 21).

Solidification of hazardous wastes can be accomplished through encapsulation. In encapsulation, the solidifying agent surrounds the waste with a compound such as Portland cement (Sell, p. 45). An advantage of this type of treatment technology is the reduced potential for leaching (Rich, p. 6–2). The disadvantages of this treatment, however, are that there is essentially no change in the toxic properties of the waste, and long-term effectiveness is uncertain (Rich, p. 6–1).

BIBLIOGRAPHY

Bretherick, L. *Handbook of Chemical Reactive Hazards* (Boston, MA: Butterworths, 1985).

Cheremisinoff, P. N. *Waste Incineration Pocket Handbook* Northbrook, IL: Pudvan Publishing, 1987).

Coffey, R. and R. Tankersley. "Company Profile Technology for Preparation and Disposal of Lab Packs," *Hazardous Materials and Waste Management Magazine* (September-October 1986).

"Exceptions for Shipment of Waste Materials," *Code of Federal Regulations*, 49 CFR 173.12, U.S. Government Printing Office (1987).

Federal Register 53(68):11744–11746 (1988).

Federal Register 51(216):40585 (1986).

Federal Register 51(247):46824 (1986).

Fennelly, P. F., M. McCabe, J. M. Hall, M. F. Kozik, and M. P. Hoyt, "Environmental Characterization of Disposal of Waste Oils by Combustion in Small Commercial Boilers," Report GCA/TR/8372-G, EPA/600/2-84/150 (1954).

"Hazardous Waste Comes in Small Packages," *Chemical Engineering Progress* (February 1989).

Kalnes, T. N., and R. B. James. "Hydrogenation and Recycle of Organic Waste Streams," *Environmental Progress* 7(3) (1988).

"Lab Packs," *Hazardous Waste Management Magazine* (November-December 1988).

Meid, J. H. *Explosive and Toxic Hazardous Materials* (Encino, CA: Glencoe Publishing, 1970).

Newton, J. *A RCRA Generator's Compliance Program* (Northbrook, IL: Pudvan Publishing, 1988).

Pettit, C. L. "EPA Acts on the First Third," *Waste Age* (July 1988).

National Research Council. *Prudent Practices for Handling Hazardous Chemicals in Laboratories* (Washington, DC: National Academy Press, 1981).

"RCRA and Laboratories" (pamphlet) The American Chemical Society, Office of Federal Regulatory Programs, 1155 16th Street NW, Washington, DC 20036.

Rich, G., and K. Cherry. *Hazardous Waste Treatment Technologies* (Northbrook, IL: Pudvan Publishing, 1987).

Sabatino, J. "Lab Pack Alternatives," *Hazardous Waste Management Magazine* (February 1989).

Sell, N. J. "Solidifiers for Hazardous Waste Disposal," *Pollution Engineering* (August 1988).

Toxic and Hazardous Industrial Chemicals Safety Manual (Tokyo, Japan: International Technical Information Institute, 1985).

Winslow, G. "First Third Land Disposal Restrictions: Consequences to Generators," *Hazardous Waste Management Magazine* (November-December 1988).

Winton, J. M., and L. A. Rich. "Hazardous Waste Management: Putting Solutions into Place," *Chemical Week* (24 August 1988).

CHAPTER 14

Regional Differences in Laboratory Waste Disposal Practices

Russell W. Phifer

INTRODUCTION TO LABORATORY WASTE DISPOSAL

Laboratory wastes have long represented a disposal problem for waste management facilities as well as laboratories themselves; the wide variety of different materials generated by laboratories cannot (and never will) be simplified or classified easily. While representing less than 1% of all chemical wastes generated, laboratory wastes cause over half the headaches as disposal facilities may well acknowledge.

Laboratory waste has traditionally been disposed of in lab packs, where small containers are packed together in a drum according to hazard class, with an absorbent material added for cushioning. This method is expensive, time-consuming, labor intensive, and yet actually fairly simple. It does present potential problems for disposal facilities, however. Improper packaging or labeling of laboratory chemicals has been responsible for accidents at several commercial incinerators, as it is difficult to check each drum for accuracy.

Nonetheless, lab packs are still the most common packaging unit for laboratory waste, as they allow commingling of compatible materials and generally do not require any analysis prior to disposal. The ultimate disposal technologies, however, have changed significantly in the last few years. As recently as 1984, nearly all lab packs were landfilled. As long-term liability issues have become more important, however, far more lab packs are being incinerated or treated.

While it has been common for a number of years for waste flammable solvents from the laboratory to be bulked and incinerated, landfilling of other laboratory wastes remained popular because of the high cost of incinerating solids and the lack of permitted facilities. The problems associated with bulking other items prevented treatment/stabilization from becoming popular as long as landfill disposal of lab packs was relatively inexpensive. Aside from the need for technically skilled labor and processing equipment, the inherent risks of combining materials were a deterrent. With landfill costs rising steadily and liability a great concern, these technologies have become more popular; incineration is now the technology of choice among nearly all large companies, primarily because of the virtual elimination of long-term liability.

While the increase in disposal of laboratory waste by incineration should surprise no one, the large amount of waste that is now being "stabilized" or treated was not anticipated. As landfill costs have risen, the wasted space in lab packs (while hold approximately 15 gal. per 55-gal. drum) has come under closer scrutiny. By mixing compatible materials in bulk, the actual volume of waste per drum can effectively be tripled. With cost to the generator of over $250/drum not unusual, waste management companies can obviously see the profit available by combining three drums of waste into one. The minuses of this method are: (1) safety concerns; (2) technical labor requirements; (3) additional material handling costs; and (4) disposal of empty containers. Nonetheless, profit motives can overcome many obstacles.

Incineration remains the most likely technology for the future of laboratory waste disposal. The most significant advantage is permanent disposal in a timely fashion; some facilities are able to process waste the same day it arrives. A growing variation on traditional incineration is the conversion of ash from the process to an aggregate for use in construction of roads. This may or may not be considered "recycling"; the jury is still out on that issue, and it will not likely be decided until Subpart X regulations are released by the EPA. For the time being, at least one facility has avoided the need for a RCRA permit under this loophole. An educated guess is that these facilities will soon be regulated by the EPA.

One key advantage of incineration is the ability to dispose of highly reactive materials after relatively simple dilution or stabilization. Picric acid is one example of a material that has traditionally presented a disposal problem (detonation has been the technology most often utilized). Most incineration facilities will accept this material once it has been diluted to a complete solution. Landfills cannot accept reactives, and other treatment technologies are expensive.

Regulatory attitudes toward alternatives to landfilling have been favorable. Despite the risks involved in both incineration and treatment, these methods are receiving considerable support among regulators. Treatment has the additional advantage of constituting waste reduction. Referred to sometimes as "trash and bash," this method is largely regional. The midwest has the largest number of facilities and companies doing this type of processing, though more and more east coast companies are utilizing the technology.

CASE STUDIES

California

Despite the large quantities of laboratory waste generated in California, the far west has the fewest facilities for disposal of any region in the United States.

There are no commercially available incinerators for lab packs and no treatment facilities accepting lab packs. Those generators that want their wastes incinerated must pay the price to transport to either Illinois or Texas, the closest states with available facilities. With transportation costs around the country averaging $3 per loaded mile, transportation can represent a large percentage of overall waste management costs. The strong influence of environmentalists and a tough regulatory stance on air quality are the main reasons for a lack of facilities. Despite the improving quality and availability of advanced equipment, it is unlikely, given these factors, that facilities will be built any time soon. The one facility that handles the disposal of reactives has a tenuous permitting situation at best, and treatment technologies have yet to take hold.

Two large commercial landfills in the state (and one in Utah) are the only disposal facilities available in the region. The distance to commercially available incinerators adds approximately $65 per drum to the cost of disposal. Still, large generators, in particular, appear willing to pay the price. The reputation of California's voters would tend to indicate they do not want facilities in the state; waste management companies are thus reluctant to spend the money necessary for engineering and permitting of new facilities without being certain of their ability to obtain an operating permit. Mobile technologies may eventually be a partial solution to the state's facility problems; for the time being, industry pays the price.

Illinois

As an industrial state with many available disposal facilities in the region, generators in Illinois have many options for disposal of laboratory waste. There are several commercial incinerators in the state, and a number of processing/treatment facilities in the region. This wealth of options and facilities has resulted in an extremely competitive situation; disposal costs are lower than in most other regions. There is a fair amount of on-site bulking of laboratory wastes, particularly among larger generators. It is likely that less laboratory waste is sent directly to landfills than anywhere else, as well. There are at least eight processing/treatment facilities in the region that stabilize or treat wastes prior to landfilling.

Michigan

Michigan is mentioned here as a case study chiefly because it has some of the most unique procedures anywhere. "Michigan rules" allow, for instance, a generator to manifest laboratory waste directly to a landfill, yet send it first to a processing facility. This facility can bulk the waste remove recyclable material, change the Manifest (listing the facility as second transporter), and send waste on to the landfill. While the "cradle-to-grave" concept would appear to be compromised, the state continues to allow this practice, apparently under a grandfather clause in the state regulations. No other state will allow interim transfer facilities such freedom.

Michigan has no commercially available incinerators, but there are several landfills in the state. In addition, there are incinerators available in nearby states, so transportation costs are minimal. Overall, disposal costs are relatively low, and generators enjoy many management options for their wastes. On-site processing of laboratory wastes (bulking) is fairly common; this can obviously result in generators enjoying reduced disposal costs, instead of allowing waste management companies to absorb more profit as in other states.

Pennsylvania

Pennsylvania is probably the most "average" of all of the case study states. Despite having no commercial landfills or incinerators that can handle lab packs, costs are relatively low due to the relative proximity of facilities in surrounding states. This is fortunate, because Pennsylvania has arguably done more than any other state to prevent construction of facilities, including California. State approval processes for waste streams to be handled in-state make life difficult for both generators and potential disposers. Any industrial waste stream, hazardous or nonhazardous, requires approximately $1800 of analytical and engineering work prior to approval by the state. As a result, it easier to send most wastes out of state. Transportation costs are reasonable because of facility locations in eastern and mideastern states. There are incinerators, for instance, in Ohio and New Jersey, and a number of landfills within a reasonable distance. One other problem in Pennsylvania, however, is that the cost of a transporter permit is extremely high. As a result, many national transporters are not permitted in the state.

There are a few waste treatment facilities in the state that accept laboratory wastes, and some on-site processing is being done. Nonetheless, as other states become aware of Pennsylvania's policies, generators in the state may begin to see problems in finding facilities for their wastes.

South Carolina

Like Pennsylvania, South Carolina is interesting because of the involvement of the state environmental agency in the approval process. South Carolina offers at least one facility for each prominent disposal technology; the state is well aware that it has, in some respects, become a dumping ground for generators in other states. It has responded by getting more actively involved in the approval process. Each waste stream must be submitted to both the proposed facility and the state for review; the state requires a minimum of 15 days to review each waste stream. Rumor has it that certain politicians in the state would like to see most of the facilities in the state closed. In the meantime, generators in the area can choose from among all of the available options.

There is even a facility in the state for "exotic" wastes such as explosives and gas cylinders. South Carolina is probably the state most able to completely handle its own waste management problems.

SUMMARY

There are significant differences around the country in how laboratory wastes are handled. In some cases, the lack of facilities has resulted in significantly higher costs. In other states, the availability of facilities in the region have enabled costs to be kept down. Differences in regulatory interpretation can be a factor, and differences in approval procedures and policies can also have an impact. It will be interesting to see how state policies change in the next few years. The states with facilities may not always be so generous in accepting out-of-state wastes, and this may make for some political battles.

Contributors

Margaret-Ann Armour, University of Alberta, Edmonton, Canada

Ralph Foster, Illinois EPA, Springfield, Illinois

Sharon Ward Harless, U.S. EPA, Philadelphia, Pennsylvania

Iclal S. Hartman, Simmons College, Boston, Massachusetts

Daniel L. Holcomb, Lab Safety Supply, Inc., Janesville, Wisconsin

James A. Kaufman, Curry College, Milton, Massachusetts

Thomas Kelley, Tufts University, Boston, Massachusetts

Joyce A. Kilby, University of Wisconsin-Whitewater, Whitewater, Wisconsin

Jennifer M. Kinsler, Lab Safety Supply, Inc., Janesville, Wisconsin

Stephen R. Larson, Massachusetts General Hospital, Boston, Massachusetts

Robert Laughlin, Canadian Waste Materials Exchange, ORTECH International, Mississauga, Ontario, Canada

George Lunn, National Cancer Institute, Frederick Cancer Research Facility, Frederick, Maryland

James P. O'Brien, Illinois EPA, Springfield, Illinois

Russell W. Phifer, Compliance Services, Inc., Wayne, Pennsylvania

Peter A. Reinhardt, University of Wisconsin-Madison, Madison, Wisconsin

Eric B. Sansone, National Cancer Institute, Frederick Cancer Research Facility, Frederick, Maryland

Max A. Taylor, Bradley University, Peoria, Illinois

Linda Varangu, Ontario Waste Exchange, ORTECH International, Mississauga, Ontario, Canada

Walter J. Warner, Jr., The Gunnery School, Washington, Connecticut

Antony C. Wilbraham, Southern Illinois University at Edwardsville, Edwardsville, Illinois

Index